# In the Deserts of This Earth

# In the Deserts

Translated from the Germ
by Richard and Clara Winst

A Harvest/HBJ Book  /  A Helen and Kurt Wolff Book  HBJ

# of This Earth

## UWE GEORGE

Harcourt Brace Jovanovich / New York and London

Printed in the United States of America

Library of Congress Cataloging in Publication Data

George, Uwe, 1940–
    In the deserts of this Earth.

    (A Harvest/HBJ book)
    Translation of In den Wüsten dieser Erde.
    "A Helen and Kurt Wolff book."
    1. Desert biology. 2. Deserts. I. Title.
QH88.G4613   1979       500.9'1'54       78-23572
ISBN 0-15-644435-6

First Harvest/HBJ edition 1979

A B C D E F G H I J

To my wife

who has been my companion

all through my years

of wandering in the deserts

# Contents

# In the Deserts of This Earth

# 1 / The Merciless Sun

## Formless and Noiseless

Sand as far as the eye can see, shimmering heat waves, no water, and not a trace of life—that is how most people imagine the desert. For those who live in well-watered, fruitful climatic zones, the desert is the ultimate symbol of barrenness and hostility to life. For many, the very idea of the desert is terrifying. They think of a boundless region without landmarks, full of perils. Especially in city dwellers, the concept of oceanic expanses of sand arouses feelings of utter loneliness. In some Arab countries the desert is called *bahr belà mà,* which means something like "sea without water." In ancient Arabic travel narratives, crossing the Sahara by caravan is often compared to voyaging across the ocean, and the southern margin of the Sahara, the Sahel zone, where the Great Desert gradually passes into savanna, is hailed as the shore toward which all voyagers must press.

Many persons who have had the experience of traveling across a desert report constant feelings of anxiety and forlornness during the journey. Many are city dwellers, and what they miss in the desert is social contact. Even the minimal contact of seeing others, of knowing that a neighbor's house is nearby, that the walls and roofs of other homes are close, is absent under the glittering, cloudless sky. They feel exposed to unknown dangers. Such experiences, and the fact that up to a few decades ago many deserts were among the least explored regions on our earth, have led to a singular use of the word *desert* in most languages. It always means the opposite of home, shelter, safety. Once, on an expedition through the central Sahara, I myself was gripped by intense feelings of anxiety for no apparent reason.

Three of us, in a four-wheel-drive vehicle, were crossing a segment of desert deep in the interior of the Sahara. In this region there was not the slightest elevation from horizon to horizon. There were no signs of any living creature, nor could the smallest plant be seen. The desert was as flat as a table, and this in an area about twice the size of New York State. Such regions in the Sahara are called serir.

Late in the afternoon we decided to stop and pitch our camp at a spot that was almost unidentifiable. Except for the abstract longitude and latitude, there was nothing to differentiate it from any other place in the landscape.

Next morning, a cool winter day, I set out—armed with canteen and compass—for a walk by myself, while my companions stayed at the camp to make notes on our observations of the preceding days. I wanted to search the rocky surface of the desert for microörganisms, the only forms of life that might possibly exist in such a place. Gradually I wandered farther and farther from the artificial landmark of car and tent. My shoes made inch-deep prints in the soil, which consisted of innumerable pea-sized pebbles. This track was probably the only sizable and visible change that had occurred in the area for the past several thousand years. I walked on toward the east. I do not know how long I had been walking when I decided to pause for a rest and looked around. A start of fright passed through me. The camp, which a short while ago had still been in sight, had vanished. That told me I must have covered more than 5 miles, the distance at which, to an observer standing on level ground or on a beach looking out to sea, the curvature of the earth becomes apparent. The curvature is what causes low landmarks and ships at sea to disappear beyond the horizon.

I was standing there wholly alone, my body the center of a vast empty disk. The horizon around me formed one uninterrupted circle. The sky was a glaring, colorless brightness, with not a cloud to be seen. Aside from me and the ground on which I stood, there was nothing but the brilliantly white, shimmering disk of the sun. But that was an astronomical distance away and was not a point of reference that had any relationship to my personal existence.

An indescribable sense of loneliness and forsakenness overpowered me. Suddenly I no longer seemed to have any sense of spatial or temporal dimensions. Here, where there were no visible standards of measurement, I felt as if I had lost all the inner standards that gave me an awareness of time and place. An infinite distance appeared to separate me from the nearest living being, and it would take me forever, it seemed, to cover that distance.

Undoubtedly these impressions were also provoked by the utter stillness around me. On that day not the slightest sound was to be heard. There was no wind to rustle among the pebbles or in my clothing. My hearing center apparently was trying to compensate for the noiselessness, because bodily functions that I had never been aware of as sounds suddenly seemed to ring loudly in my ears. I heard every heartbeat and even the coursing of the blood in my veins. A swallow from my canteen produced a loud gurgling noise as the water ran down my gullet into my stomach. This condition of oversensitivity rapidly mounted and became anxiety. I felt as if I were the sole living being on a remote, lifeless planet. My glance fell upon my track, which instantly assumed enormous importance. It was proof that I was a being with a past and a future, and the way it ran to the rim of the disk, to the horizon, showed me that my walk must have had a beginning. I began running back in my own track, and ended my terrified dash only when the camp reappeared as a tiny point on the horizon.

That night I thought for a long time about the astronauts who first set foot on the moon, and I realized how sensible it is for men to undertake such adventures in pairs. The pair is surely the smallest unit in which the more highly developed life forms can endure cosmic dimensions. I also know, from that experience, that total noiselessness except for the sounds of one's own body can exist only in the heart of a great desert. Ever since, even in soundproof rooms, I have been able to hear the faint residual noises that penetrate from outside through the insulation.

Instinctive anxieties stirred by the incomprehensible expanses of the desert are only one aspect of the range of human response. Another aspect is a sense of boundless freedom. Our environment has become so complicated that the individual is no longer able to grasp the whole intuitively and naturally. All of us feel that our freedom of action as individuals has been circumscribed to some degree; many of us feel that largely anonymous powers govern our lives.

The experience of being dependent on ourselves alone as we tramp through the boundless expanses of the desert, of trusting only our own abilities, enables us to recapture our identity. We can shake off the dross that our production-oriented society has heaped up in our psyches and return to the dreams of childhood. In the desert, discoveries can still be made—and what a host of mysteries might be hidden in those vast expanses! Seen in this light, the desert is a landscape in which the human spirit can exult, a landscape that can satisfy man's deepest craving, which is for spiritual freedom. A French Sahara explorer once said, "The desert is probably the most intensely loved landscape on our earth."

But there is still another aspect of the desert: the cruel, merciless side that is expressed in famine. In recent years the deserts of the globe have been expanding exponentially. Every day the area of desert grows by almost 40 square miles, and every day some 200 persons die from the direct or indirect consequences of the expansion. The people most affected are those who live on the fringes of the Sahara. Here the climate is steadily growing more arid. Rainfall ceases entirely for many years on end, causing the springs to dry up—those indispensable springs that supply water for drinking and for irrigation of the fields. Sandstorms occur with increasing frequency, devastating the growing crops and choking the last springs with sand. The natural vegetation of bushes and grass becomes parched. The herds of goats, sheep, and camels, the pride and wealth of the nomads, can find no nourishment. The animals die of hunger and thirst. Ever more acute food shortages lead to undernourishment among the people. Their resistance lowered, they fall a helpless prey to disease. The frightful famines that have visited Africa, especially those of recent years, are only the ghastly prelude to a series of famines that will descend on humanity in the next several decades. Africa is a drying continent across which the desert is continuing its inexorable advance. People are being crowded into smaller and smaller areas, partly because of the steady expansion of the desert, partly because of a galloping birth rate. The desert will ultimately win the grim race. It is the growing desert that will in the end prevent the limitless overpopulation of this planet. The frequently cited and dreaded number of 20 billion human beings by the end of the present millennium will in all probability never be attained.

## A Vast Unexplored Region

The Sahara, the greatest of them all, was the first desert to cast its spell on European travelers, chiefly scientists. Situated at the very gates of Europe, it could be reached fairly easily. And to the present day the Sahara has remained the very essence of what we mean by desert. The Romans were probably the first "foreigners" to obtain any detailed knowledge of the area. As early as 2,000 years ago, they undertook numerous expeditions into the Sahara, and it is likely that they were the first to succeed in crossing it from north to south. Unfortunately, the accounts of the Roman geographers were lost or largely ignored.

Arab scientists of a thousand years ago had a good deal of exact knowledge about many parts of the Sahara. But, because their language and script were relatively unknown in Europe, their knowledge remained largely inaccessible to the West. Eighteenth-century maps generally showed a large blank where the Sahara lies. Occasionally the blank would be filled with drawings of fantastic creatures that spat fire or poison and devoured interlopers. Even a hundred years later, in the middle of the nineteenth century, the great naturalist Alexander von Humboldt thought that the Sahara was one gigantic, monotonous wasteland of sand. Astonishingly enough, that same totally unrealistic picture of the region persists today in the minds of many. If we took a survey, it probably would be found that ninety-five out of a hundred respondents held that view.

In 1826, the Scotsman Gordon Laing became the first European to reach the legendary desert city of Timbuktu. A year later René Caillie also reached that important caravan crossing on the southern margin of the Sahara. A period of systematic research opened in 1850, when the German Hermann Barth undertook a five-year exploration of North Africa. He was followed in 1863 by Gustav Nachtigal, who was the first to visit the 11,000-foot-high Tibesti Mountains of the eastern Sahara. Nachtigal later traveled on into black Africa, to what was then the Bomu Kingdom. Nowadays it is almost impossible to conceive the hardship and peril to which those first explorers were exposed. They were entirely on their own for years, traveling in regions where no European had ever set foot and where conditions often were so dangerous that even native guides would refuse to go along. The Christian "unbelievers" frequently had to disguise themselves as Arabs and travel with caravans through searing hot, waterless wastes in order to reach the goal of their explorations. To avoid betraying themselves, they had to have an accurate knowledge of the language and customs of the various tribes. In many cases, disclosure would have meant certain death.

When Gustav Nachtigal returned to Germany a few years later, he reported on his adventures directly to the king of Prussia, who was so enraptured that he asked the explorer to take two life-size paintings, of himself and his queen, to the king of Bomu. Nachtigal accepted the mission and set out once more on years of dangerous travel across the Sahara. On July 5, 1860, he presented Sheik Omar of Bomu with the portraits of the Prussian monarchs. The books and articles that Barth and Nachtigal published about their travels were best sellers of the period. Today they languish unread in the basements of libraries.

There was also a woman among the first explorers of the Sahara, an Englishwoman with the French name of Alexandrine Tinné. According to contemporary accounts, she was a striking beauty. In 1869— she was then just twenty-nine years old—she attempted to cross the Sahara by caravan in order to reach the interior of Africa. In the vicinity of the Libyan oasis of Murzuk, the caravan was attacked by Tuaregs and completely plundered. Miss Tinné was taken captive, and subsequently her right arm was hacked off. She bled to death.

The establishment of the colonial empires coincides with this early period of European exploration. Between 1830 and 1880 the French conquered large parts of North Africa. Thereafter, desert research was conducted mainly by the military. A large number of fortified desert posts were set up, and scientists were able to use them as bases. But let us turn to more recent times.

Any account of modern exploration and study of deserts must give prominence to the name of Théodore Monod. This French savant is often considered the foremost expert on the Sahara. From 1934 on, he aroused great excitement by his daring camel rides through the remotest and least-known parts of the Sahara. On December 12, 1954, Monod set out from the small Mauritanian oasis of Waddan, in the western Sahara, with two natives and several riding and pack camels. His aim was to investigate one of the last unexplored regions. For weeks he traveled in the no man's land of the southwest Sahara, during the daytime lashed by fiery sandstorms, at night exposed to biting cold. Frequently Monod and his companions walked all day in order to spare their riding camels. At times they were so exhausted that they held on to the tails of the camels and let themselves be dragged along. The small caravan traversed a totally waterless region of some 550 miles, incidentally setting a world's record. Never before had anyone covered such a great distance between water holes in the desert.

In the course of his journey Monod made some sensational discoveries. He found vast dried lakes in the interior of the Sahara, with the remains of ancient fishing villages along their shores. Thousands of bone harpoons lay strewn about in the desert sand. The dried lake bottoms were covered with the bones of hippopotamuses, fish, and crocodiles. (We shall say more about this mysterious find in a later chapter.) After twenty-two days, the caravan reached the Arauan spring. The travelers refilled all their vessels with fresh water and set out again the same day. Once more they covered over 500 miles, through a region represented on their maps as a huge blank area. After a march that had lasted one and a half months, they arrived, weary and emaciated, at the oasis of El Galluya. All in all, they had covered approximately 1,050 miles.

Today some desert exploration is conducted by aerial photography, from planes and satellites. But the main work still must be done on the ground, and unexplored areas remain. There are regions of the eastern Sahara in which no one has set foot to this day. However, modern desert research is no longer so bent on filling in the last blank spaces on the maps. Rather, science is concerned with investigating the desert's function in the earth's history and its effect on the evolution of life forms. The difficult and complex questions being raised can be answered only by interdisciplinary endeavors, and biologists, geologists, paleontologists, climatologists, and others are working together in desert research. A science that could comprehend and describe the phenomenon of deserts would possess the key to the past and future of our planet.

## The Causes of Desert Formation

Theoretically, any region on the globe could become a desert if no clouds formed above it for a considerable length of time. No rain would fall, and the sun's rays would blaze unhindered on the surface of the soil. All vegetation would dry up, and in a short while the area would be transformed into desert. Science has developed criteria for distinguishing genuine deserts from landscapes of a similar appearance—for example, regions of bare rock in mountains or the type of barren that is produced by overgrazing. Since rain is one of the prerequisites for life on the continents of our planet, the biological forms that prevail in any given region are dependent on the level of precipitation. Another vital factor is the specific amount of solar radiation. We know what climatic conditions match each type of landscape on our planet. In addition to the major climatic zones, such as the tropics, the subtropics, the desert, the arctic, there are innumerable miniclimates that deviate considerably from the overall climate of the zone in which they lie. The kinds of plants that grow on northern slopes, for example, differ from those on sunny southern slopes. Anyone who has ever tended a garden or house plants knows something about these differences. Some plants will grow only under large trees in damp, shady miniclimates; others need a sunny, dry spot if they are to thrive. A typical miniclimate zone within a desert is the oasis. Oases do not receive any special dispensation of rainfall; their green growth is due to the presence of ground water.

An area is classed as a desert if it receives less than 10 inches of precipitation annually. In many desert regions, especially in interior zones, the rainfall is far less than that. The lowest average for any region on our earth is a mere 0.4 inch per year. By contrast, tropical rain forests receive an average of nearly 80 inches of precipitation annually, and some are drenched by almost 200 inches per year. While other climatic regions have definite rainy and dry periods, rain falls on the desert with extreme irregularity. There is hardly a desert region on earth where the average 10 inches of rain falls consistently every year. In the heart of the Sahara are regions where not a drop of rain has been registered for over twenty years, although the annual average there is 4 inches.

Then someday a mass of wandering clouds happens to drift into the region and abruptly deposits an inconceivable amount of rain. Anyone who has experienced such a rainfall in a desert has the feeling that some gigantic sluice has been opened. Individual drops of rain cannot be distinguished; the water seems to fall from the sky in a vast sheet. It immediately produces great floods on the surface of the desert, which has no plant cover to protect it. The water finds its way into hollows, gouging out deep channels as it moves. That explains why, even in the driest regions of a desert, the traveler everywhere encounters traces of running water. The wadis, as the dried brooks and riverbeds of North Africa are called, fill up with remarkable speed and become raging torrents. The force of the water is so great that the yellowish brown flood carries thousands of tons of mud, sand, and scree along with it. You can be standing on the rim of a riverbed that has been parched for decades, and in the stillness of the desert you suddenly hear the sound of a distant humming. Within minutes the hum rises to a gurgling thunder, and you see a wall of whirling brown water, higher than a tall man, pouring around a bend in the wadi. Before you have recovered from your first start of alarm, you are watching a flood that can roll large rocks before it. Rare though such cloudbursts are, perhaps more persons have been drowned in the desert than have died of thirst. One of the basic rules of desert travel is: never camp in a wadi. But the rivers usually flow for only a few hours, and they almost never reach the distant sea. After a short course they trickle away into the subsoil.

Within a few hours, then, an amount of rain can fall that, according to the meteorological statistics, should be the desert's proper ration for decades. And yet such wild storms are reckoned in the averages.

The lack of rain and the high temperature of the desert are closely connected. Since clouds form so rarely and the moisture content of the air is very low—often between 2 and 5 per cent—the sun's rays strike the desert floor unhindered. Up to 95 per cent of the solar radiation reaches the ground. In temperate zones, when the sun's rays reach the surface, radiation heat has been reduced already by about 60 per cent; the rest is absorbed by clouds, by plants, by air humidity, and the water of lakes and rivers. Again the tropical rain forest represents the extreme contrast to the desert. Because of clouds, very high atmospheric moisture, and the dense ceiling formed by the leaves of the jungle trees, sometimes only 1 per cent of the sunlight reaches the floor of a rain forest.

In the desert, the intense solar radiation produces a tremendous heating of the ground. In areas of the Sahara where the surface consists of black rock, ground temperatures of more than 85° C. (185° F.) have been recorded. During the summer months it is impossible to enter such an area on foot. Nevertheless, these hottest and driest places on the surface of the earth offer a home to a few plants and animals, as we shall see later in this book.

A peculiar phenomenon of many deserts is what is called a "ghost rain." The few clouds that have managed to float over a desert region suddenly condense. Like a black curtain, rain begins to fall. Then, shortly before it touches the ground and the thirsting plants, the rain encounters a layer of heated air that has formed just off the ground. Instantly it evaporates, and most of the already rare rainfall is lost. The cooling produced by the evaporation sometimes causes the layer of hot air to dissipate, so that if there is a second downpour it has a chance to reach the ground.

The same atmospheric conditions that produce a high degree of solar radiation by day lead to a rapid cooling at night. The absence of clouds and the low level of atmospheric moisture make it possible for 95 per cent of the heat stored in the ground to be radiated away and to escape into the higher strata of the atmosphere. In temperate climates, on the other hand, only about half the warmth stored in the ground is radiated away at night; the other half is radiated back to the ground by clouds and vegetation.

The distribution of deserts on our planet is by no means a matter of chance. Most of them lie in the subtropics, or horse latitudes, the belts along the Tropic of Cancer and the Tropic of Capricorn. Two such belts of desert circle the globe, one northern and one southern. Among the deserts of the northern subtropical region are the Sahara, the desert of the Arabian peninsula, and a portion of the desert of Iran, as well as the deserts of Mexico and the southern part of the North American deserts. In the region of the Tropic of Capricorn lie the Kalahari Desert of South Africa, part of the Patagonian steppe in Argentina, and the continent of Australia, much of which consists of desert and parched steppe.

Since the continental masses of the Northern Hemisphere are much larger than those of the Southern Hemisphere, the deserts in the northern half of the globe are correspondingly larger. The reason for the formation of the subtropical deserts lies in the spherical shape of the earth and the inclination of its axis of rotation, which is 23.5 degrees. As the earth revolves around the sun once every 365 days, the axis of rotation, small variations aside, maintains that inclination. The inclination causes the northern half of the earth to be turned toward the sun part of the time. In summer the sun stands at its zenith—vertically—over the northern tropic, and the seasonal heating of the earth's surface is greatest in the latitudes of that tropic, since the factor that determines the annual temperature fluctuations on earth is not the distance the sun's rays must traverse, but the angle at which they strike the surface. Because the earth's axis of rotation does not change, after six months the Northern Hemisphere is turned away from the sun. The sun then stands vertically above the southern tropic, the Tropic of Capricorn, and produces its maximum heating effect there. Over the course of a year the solar zenith changes from the northern to the southern tropic and back again. In this transit, it is the region between the tropics that is heated the most. That is the principal reason why most of our deserts lie in the subtropics.

Another factor in the formation of deserts is variations in air pressure around the globe. The variations occur because some air masses are warmed more than others and also because the rotation of the earth imparts movement to the atmosphere in the form of powerful vertical and horizontal currents. For example, everything along the Equator is moving at a speed of more than 1,000 miles per hour. The strongly heated air masses over the equatorial region generally rise.

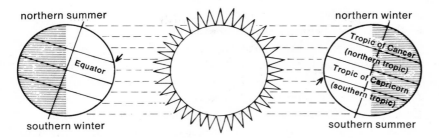

northern summer                                    northern winter

Equator

Tropic of Cancer
(northern tropic)
Tropic of Capricorn
(southern tropic)

southern winter                                    southern summer

Annual revolution of the earth around the sun

The result is a zone of low atmospheric pressure. The rising air masses cool off and lose all their moisture by condensation—one of the reasons for the heavy rainfall in tropical zones. The dry air masses then descend upon the subtropical regions on either side of the Equator, producing zones of high pressure. As they descend, they warm up and absorb whatever moisture is present in the lower layers of the atmosphere and on the ground. The consequence is dryness in these parts of the globe. North and south of the subtropical regions we again find zones of generally rising air and low pressure. Over the polar regions the air masses drop back down, and high pressure and dryness prevail. Fundamentally, the polar regions of the earth are icy deserts. Most people find this baffling, because the polar deserts are so extremely cold and we tend to think of heat as the prime characteristic of a desert. On the whole that is true, but there are exceptions. Elsewhere in this book we will discuss deserts that are among the coldest regions on earth—temperatures there drop to −50° C. Nevertheless, the polar deserts are somewhat friendlier to life than the hot deserts are because their small amount of precipitation does not evaporate immediately. Most of the precipitation falls as snow, which accumulates over the long polar winter. During the brief summer, when some of the snow and ice melts, plants and animals can obtain the water they need, and a brief phase of lush development of living forms follows.

The earth's rotation also moves the enormous masses of water in the oceans and is responsible for the vast circular currents of the world's seas. Paradoxical as it may seem, these great ocean streams are a further important factor in the genesis of deserts. Currents that originate in the polar seas transport immense volumes of cold water toward the Equator. Most of these cold currents pass along the western side of the continents, or, more specifically, the edge of the continental shelves. The rise in the ocean bed forces the cold streams to the surface. Winds blowing across the oceans toward the land masses cross these cold currents, which cool the air so that it is unable to hold large amounts of moisture. Coastal regions rimmed by such cold currents become deserts even when they are located in the tropics. The best known of the coastal deserts lie in the Southern Hemisphere. The Benguela Current, which flows out of the Antarctic Ocean, has produced the Namib Desert on the coast of South-West Africa. Another coastal desert, which runs for more than 1,200 miles along the Pacific coast of South America, is the Atacama Desert. Its formation is due to the Humboldt Current, which prevents rain clouds from reaching the coast. The seaward edge of the Atacama Desert represents an area as hostile to life as any place on our planet; in this respect it resembles large areas in the interior of the Sahara, Death Valley in the United States, and the Dasht-e-Lūt Desert of Iran. Vast parts of these regions have known no rain whatsoever for many decades. If we stand on the Pacific coast of South America and gaze out to sea, we see millions upon millions of sea birds hunting the swarms of fish that inhabit the cold waters of the Humboldt Current. If we direct our gaze toward the interior, we see total desert. The weatherworn mountains are covered with a dense layer of gray dust. It is an almost sterile landscape, without the slightest sign of life.

Coastal deserts also exist in the Northern Hemisphere. Mexico's Lower California peninsula receives rain only once every five or six years. The cold California Current that comes down from the Arctic prevents moist air masses from reaching the peninsula.

In addition to subtropical deserts, arctic zones of low precipitation, and coastal deserts, there are relief deserts. As the name suggests, this type of desert is related to certain areas on the relief map of the earth's surface. Relief deserts form in areas beyond high mountain ranges or on high plateaus that rain clouds cannot reach. In rising, the clouds cool off, and their rain falls on the outer slope of the mountain range or plateau.

Extensive regions in the western half of North America are relief deserts. High mountain ranges prevent clouds from the Pacific from reaching the interior. Similar conditions prevail on the high plateaus in Iran, southern Russia, and Afghanistan: mountains bar the way to clouds from the Indian Ocean.

But the largest relief deserts are to be found in the interior of Asia. They include the Takla Makan Desert, in the western part of China, and—probably the best known of the Asian relief deserts—the Gobi

Desert, which covers large areas of northern China and Mongolia. Some of the inland Asian deserts owe their formation partially to the Himalayas, the highest mountain range in the world, where altitudes reach nearly 30,000 feet. Other inland Asian deserts are the combined result of the height of nearby mountains and their great distance from the sea.

Although relief deserts as a whole make up a smaller area of the earth's surface than subtropical deserts, geologically they represent the most highly evolved type of desert. This is related to the astonishing fact that mountains and high plateaus, which produce the relief deserts, are constantly being formed anew and appear to be growing higher and higher. I shall return to this phenomenon in a later chapter. In contrast to other types of desert, which are limited to certain latitudes, practically any part of the earth's surface can become a desert because of its elevation. And a conjunction of all the desert-forming forces could someday transform the entire earth into a dry, uninhabitable wasteland.

Deserts and parched steppes today cover a large part of the continental area of the world. If all the dry zones, including the arctic regions, are counted, we find that desert covers 30 per cent of the continental surface. Africa leads the way in dryness. Of that continent's 11.5 million square miles, approximately 4 million consist of desert. In addition, there are huge areas of dry, desertlike steppes

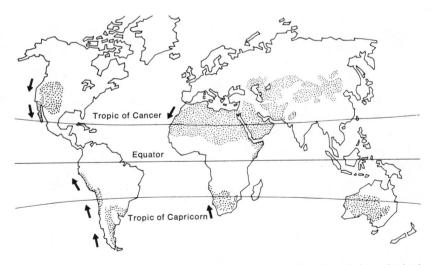

Map of the world's deserts. The arrows indicate the direction of the principal cold ocean currents that affect the formation of coastal deserts.

and savannas that are sometimes extremely arid. Altogether, Africa is about 50 per cent desert and arid regions. The Sahara, with its approximately 3.5 million square miles, in itself assumes nearly continental proportions. The size of this "desert continent" becomes even more impressive if we add to it the 1 million square miles of the desertlike Arabian peninsula. In climate, geological structure, and surface, the peninsula can quite properly be grouped with the Sahara. The two deserts were separated in comparatively recent geological times by the irruption of a new, embryonic ocean, the Red Sea. The north-south extent of this desert region is approximately 1,200 miles and the east-west distance almost 5,000. In other words, the Sahara-Arabian desert covers nearly 11.5 per cent of the land area of the globe. Such a tremendous desert can be accounted for only by a conjunction of many desert-forming forces. First of all, the Sahara is a subtropical desert. In addition, the cold Canaries current from the North Atlantic runs southward along the western coast of Africa, keeping moisture-laden winds away from the Sahara. The tall Atlas Mountains along the northern rim of Africa keep rainy low-pressure systems from the northeastern Atlantic and the Mediterranean out of the North African interior. Those three factors have combined to make a desert so large that even if enough rain clouds were available they could never traverse the vast distances to reach its heart.

## The Greatest Sandblasting Machine on Earth

That brings us to an interesting phenomenon we might call self-parching. Once a large desertlike area on our planet has achieved a considerable size, as the result of the conjunction of several factors, its continuance and expansion seem to be subject to a peculiar inherent dynamism. A simple example will clarify the point. Suppose a clearing is cut in a large forest. This creates the risk that when storms attack, their high winds will blow down more trees, since trees in the interior of a forest have less resistance to wind than those on the fringes. If the clearing is not replanted, it will tend to spread of its own accord.

A similar inherent dynamism is found wherever a large area of pasture in a dry climatic zone—such as southern Europe, for instance —develops a bare spot from overgrazing. When the protective layer of vegetation is removed, the soil is exposed to the forces of the weather. It dries out and is carried away by the wind. Gradually a hollow develops, in which the wind creates vortices that act like a drill. The erosion scar steadily enlarges. If several such spots are present, the wind transports eroding material, in the form of sand and dust, from one to the next. The sand grinds like emery. Once this self-destructive process has been initiated in a region, it tends not to stop. In the same way, the dry winds that have created the desert will enlarge it beyond its original dimensions. The notorious sandstorms of the Sahara are as greatly feared by the inhabitants of its oases as they are by travelers.

There are different types of sandstorms in the various parts of the Sahara, and they bear different names. In Libya there is the ghibli, also known as the sirocco. It blows from the south and travels as far as the Alps. In Egypt the hot and extremely dry khamsin often blows for days. Harmattan is the name for a sandstorm that occurs repeatedly during an eight-month period along the southern rim of the Sahara. This type of storm can involve an area of well over 2 million square miles. But *sandstorm* is a very general word, and many of the sandstorms dramatically described in travel narratives are not really that. Dust storms are far more frequent than sandstorms.

It is necessary at this point to differentiate between sand and dust according to the standards set by desert scientists. A grain of sand has a diameter of between 0.2 and 2 millimeters, a grain of dust a diameter ranging from a few hundredths of a millimeter to 0.2 millimeter. Since dust is so much finer than sand, it can easily be transported by the wind, and even moderate wind velocities will unleash enormous dust storms. The desert dust storms of the Sahara are among the most impressive and terrifying spectacles of nature. They appear suddenly on the horizon, in the form of a tremendous and seemingly impenetrable moving mass as much as 10,000 feet high. Travelers witnessing the onrush of one of these storms for the first time have reported that the inexorable advance of the yellowish brown wall of dust is utterly terrifying. The sight alone gives one a feeling of being crushed or of being about to suffocate.

Most such dust storms are generated during the winter months by an advancing cold front. The dust is so thick that even the glaring desert sun cannot penetrate it. In the center of the storm, visibility is absolutely zero; black night prevails in broad daylight. The desert traveler or desert dweller must cover nose and mouth with wet cloths as soon as the advancing storm is seen. Fortunately, the walls of dust are short-lived phenomena. Their diameter ranges from 6 to 12 miles, and frequently all is over in half an hour. The sun breaks through, and then the wall can be seen from behind. But the fine, light dust can remain floating in the atmosphere for a very long time after the wind has stopped—for days, months, even years. The dust floats up to altitudes of more than 30,000 feet, and the trade winds drive it around the globe at great heights, so satellite observations indicate. Through its floating dust, the Sahara probably exerts a global influence on the world's weather. A similar phenomenon was observed in 1883 when a tremendous volcanic eruption took place on the island of Krakatoa. Quantities of the volcanic dust that was spewed into the atmosphere floated around the globe at great heights for years afterward.

Another phenomenon produced by dust in the Sahara is what is called dry fog. It appears on hot days when the wind is feeble or there is no wind at all. Large amounts of very fine dust enter the atmosphere along with the rising hot air. The fine particles can spread out over wide areas of the desert and remain there for months. In a light wind they move across the desert much like swaths of mist in the mountains. Since visibility in a dry fog is only a few feet and the sun is obscured all day long, the fogs constitute a great danger for caravans and even for travelers by car. Orientation is impossible; important landmarks and springs cannot be seen.

Unlike dust storms, sandstorms form only in high winds. But many sandstorms are preceded by large dust storms, since the rising wind picks up and moves the dust particles first. As the velocity of the wind increases, the sand also starts to move. The storm almost always begins with narrow ribbons of sand moving along an inch or so above the ground. Within a short time the ribbons join, until a gigantic unbroken yellow carpet of sand is gliding along at great speed over the surface of the desert. As the velocity of the wind continues to increase, larger and larger masses of sand are blown into the air, and the top of the sand carpet continues gradually to rise. That is a sandstorm.

I have experienced several sandstorms, one of the worst of which was in the interior of the Sahara, near the oasis of In Salah in Algeria. The storm began in the manner I have just described. First came moderate winds and tremendous drifting masses of dust, which gradually changed to a dense carpet of sand hovering over the ground from horizon to horizon. The wind became stronger. It blew strongly and evenly, and soon the air seemed to consist of nothing but yellow grains of sand, which rattled against the windshild of our car as if someone were throwing shovelfuls of sand. At times visibility dropped to zero. Before everything was blotted out, I had noticed that the track I was following ascended a rise several hundred yards away. I drove straight on until I reached that hill, and the effect was like that of a submarine suddenly breaking through the surface of the water, for abruptly the car rose above the upper limit of the sandstorm. The spectacle I beheld was unique. I stood under a blazing sky and looked down at the top of a seemingly infinite sea of sand flowing along at high speed. The brow of the hill on which I stood formed an island in the midst of this surging golden yellow sea. At some distance I could make out the brows of other hills, other "islands." But the most curious part of the experience was the sight of the heads and humps of several camels that appeared to be floating on the surface of the sand, like ducks drifting in the current of a river. The camels' heads and humps, all that showed above the upper limit of the sandstorm, were moving in the same direction as the drifting sand, but a great deal more slowly.

An abrupt, sharp upper limit is typical of sandstorms, for the reason that the storms are so often formed by very steady winds. Unlike particles of dust, grains of sand are almost all of about the same size and weight. A specific wind velocity will raise the uniform grains of sand to a uniform height. How high a dense, drifting blanket of sand will reach depends in large measure on the amount of wind energy available. Very strong winds can raise the sand to a height of about six and a half feet, but no more. Dust, on the other hand, is lifted to far greater heights because of its minimal weight. Moreover, since dust particles differ so greatly in size, a dust storm does not have a sharp upper boundary. There is a gradual range of sizes, from the larger particles of dust, which are carried along near the ground, to the smallest, which may fly thousands of feet up—although the differences in size cannot be discerned by the naked eye.

If every sandstorm could be scientifically observed and recorded, the maximum height would undoubtedly prove to be the figure just given, about six and a half feet. Because of the lack of scientific observation posts in the desert, no reliable statistics are available. But there is indirect proof of the assumption that the sand whirled up by sandstorms usually reaches that height. It is the height of camels, which have become adapted to the desert. The heads of these beasts reach just above the densest zone of drifting sand, giving them a prime requisite for survival in the desert: the ability to breathe air that is relatively free of sand. The characteristic way camels carry their heads high also helps to bring their nostrils above the probable maximum height of the drifting sand. But dust is something else again, and even camels can be enveloped by a dust storm.

Desert mammals that are not as tall as camels and therefore cannot hold their heads above the blowing sand rely on complicated devices to protect their respiratory organs from penetration by sand and dust. A good example is the saiga antelope of the desertlike steppes of southern Russia, which protects its air passages by highly effective filters.

In wind velocities ranging between storm and hurricane, the sand can be whirled higher than six and a half feet. Then the masses of sand begin to whirl in the form of gigantic cylinders. One cylinder after another rolls along, sometimes continuing on for days. Since sandstorms of this severity often are caused by sinking hot air, the temperature may rise unbearably. Air temperatures as high as 58° C. (135° F.) have been measured. Simultaneously, the moisture in the air drops to only a few per cent. Desert travelers who are exposed to severe, long-lasting sandstorms are in grave peril of their lives. Even when clothed, the human body can lose up to a quart of moisture an hour from the drying effect of such a storm. Unless large quantities of drinking water are available to balance out the loss of fluids, death from dehydration can occur within a few hours. Persons have been found dead in the desert, their bodies completely dried mummies, because the entire fluid content of the body evaporated during a sandstorm.

One extremely unpleasant accompaniment of a sandstorm is the rise in atmospheric electricity. The mutual friction of the grains of sand can increase the electric potential in the air to some 80 volts per square meter. A voltage this high seriously disturbs the electrical field of the human body, resulting in unbearable headaches and nausea. The German geographer van der Esch, who, shortly before World War II, surveyed large parts of the eastern Sahara for the first time, suffered from such headaches during a sandstorm that raged for days. He hit on the idea of discharging the electric potential that had accumulated in his body by using a metal rod to conduct it into the ground. He walked around in the sandstorm employing the steel handle of his automobile jack as a cane. The method proved effective. In half an hour, he reports, his headaches perceptibly diminished.

Dust and sand are the end products of the weathering of the various rocks that compose the earth's crust and likewise make up much of the desert's surface. The great difference in temperature between day and night is one of the chief causes of weathering in the desert. Rocks unprotected by any cover of vegetation can heat up to more than 80° C. (175° F.) by day and within a few hours after sunset cool down to 10° C. (50° F.). Under the constant strain of such wide temperature fluctuations, the flanks of desert mountains crack and burst. Fragmented rocky strata slide down into the valleys and break up into smaller and smaller pieces. Successive strata of rock are exposed and subjected to the same kind of destruction. Especially during the hot summer months, a traveler in the mountains of the desert is likely to hear a constant series of reports caused by bursting stones. Gaping cracks as much as 300 feet long can form in the rock. I remember once resting near a large rock about 3 feet wide when it suddenly gave a loud report, like a cannon being fired, and shattered into several pieces. Not long before, a brief but violent downpour of rain had cooled the heated stone so abruptly that its surface steamed.

Such fracturing, which is typical of desert rocks, is not due to great temperature changes alone. In addition, there is the relief of pressure that occurs when rocks that have lain at greater depths are exposed by the weathering of the top layers. For many millions of years they have been subjected to the tremendous pressure of the strata of rock above them, and the crystal structure of the rocks has adjusted to the pressure. Exposure by weathering destroys the equilibrium; the rocks no longer have the external pressure to counter their internal stresses. A fairly small triggering action, such as the temperature difference between day and night or a sudden cooling by rain, can produce the fracturing.

Many desert rocks display what is called exfoliation. Such rocks are round stones that look like cabbage heads and are even wrapped in thin layers of stone resembling cabbage leaves, which break away easily. This is a type of weathering caused by the minerals in the stone. What little water is available penetrates the stone through fine crevices caused by temperature fluctuations. The amount of moisture is, however, insufficient to dissolve the minerals and wash them out of the rock, as is the case in wetter climatic zones. The minerals, mostly in the form of salts and clay, swell up, closing the crevices and empty spaces, and producing a strain that can amount to as much as 140 pounds per square inch. These strains cause the stones to break apart in thin layers.

All the mountain ranges and rocks of the desert are gradually shattering. The rock decay produces extensive areas that appear to be covered with nothing but rubble. Often the rock is so weathered and brittle that it crushes under the desert traveler's shoes. You feel as if you were tramping over a great expanse of broken roof tiles. The end product of the disintegration is either sand or dust, depending on the type of rock. Dust, for example, forms from solidified clay or volcanic deposits; the larger grains of sand come from the weathering of very hard rocks, such as granite, sandstone, or limestone. The sand grains, which can be discerned with the naked eye, probably represent the minimum size for these rocks. Because of the spherical shape and consistent size and weight of the grains of sand, they provide no surfaces for mutual weathering. That is why sand does not decay further into dust, but, instead, remains in that form, as vast stretches of sand dunes. The dust and sand that result from weathering, combined with the action of the wind, accelerate the process of weathering more rock. Storms act like a gigantic natural sandblasting machine. The steady desert wind hurls grains of sand against still-unweathered rock and grinds it up into sand. More sand provides additional grinding material, which in turn brings an increase in the pace of erosion. A kind of chain reaction inexorably hastens the self-destruction of the rocky landscape.

One incident will serve to demonstrate how powerful the grinding action of grains of sand can be. Several automobiles were crossing the Libyan desert when a severe sandstorm arose, with a wind velocity of hurricane strength, and lasted for several days. After forty-eight hours, all the paint on the vehicles had been stripped away by the grains of sand, and the chassis were down to their bare metal. Moreover, the glass had become opaque; the windshields had to be knocked out before the drive could continue.

The weathering process means that ever larger volumes of new sand are constantly being generated in the desert. The reduction of rock assumes gigantic proportions. Sand from disintegrated desert mountains is gradually transported by the wind and piled up again, at great distances from its point of origin, in the form of mighty mountains of sand dunes. Most people think of sand dunes as the very quintessence of the desert. But sandy areas constitute a much smaller part of the earth's total area of desert than is generally assumed. In the Sahara and Arabian desert region, nearly 1 million of the 4.5 million square miles are sandy desert, or only about 20 per cent. By far the larger part consists of desert mountains, which reach heights up to 11,000 feet, vast stony plateaus—the hammadas—low-lying dust-filled basins, those seemingly endless pebbly wastes called serir, and innumerable small lines of hills that have disintegrated into black heaps of slag and rubble.

Because of the vastness of the Sahara, however, the relatively small portion of sandy desert assumes a false proportion in relation to the other types of desert terrain. "Only" 20 per cent is still 1 million square miles, which is nearly four times the size of Texas. Several of the sandy deserts distributed here and there in the Sahara each have a coherent area four times as big as the state of New York.

The dunes look like the frozen waves of an ocean; the extent of the sandy deserts, stretching as they do from horizon to horizon, also gives an impression of oceanic vastness. The sight of the black, sun-scorched landscapes of rubble in stony or mountainous areas of the desert produces a feeling of devastation and chaos. In contrast, the look of a sandy desert gives aesthetic pleasure. Instead of jagged, broken fragments of rock, the eye beholds rounded, harmonious forms—curved mounds and gently rolling hills of golden sand. The breath of wind that ruffles the watery surface of a lake and makes the reeds gently incline their heads, or the strong wind that whips up the waves of the sea, bends trees on land, and transforms a grainfield into a single billow—all the various winds that imprint themselves for a moment on the landscape seem to create permanent images of themselves, to achieve permanent form, in the dunes of a sandy desert. All the wind's characteristics are, as it were, petrified in the sandy desert. Small ripples in sand, barely an inch high; billows of sand running every which way, some with their crests breaking over, some gently rounded; waves of sand several feet high marching in endless succession across a vast distance. . . . Instead of chaos and chance, the human eye finds a system of rhythmically recurring forms. For all that this terrain is so hostile to life, the rhythm and the delicacy of its forms produce in human beings a sense of well-being, of comfort.

Some of the Sahara's largest regions of continuous sand dunes are found in the lowest places, in enormous shallow geological basins. Winds from all quarters deposit their freight of sand in these dips, and mountains of sand result. These regions, which are called ergs in the Sahara, have some dunes 1,000 feet high: the star dunes. They are composed of ridges of dunes that rise to a peak. In contrast to other types of dunes, many of which migrate or constantly change their form, star dunes are fixed in one place and can serve for many decades as landmarks, since they are subject to little change. The great complex of ergs was created over hundreds of thousands of years, and, while the wind can work small alterations in an erg, it cannot change the region as a whole.

Sometimes the wind sweeps sand up into small movable piles, little sickle dunes, which are called barchans: migrating dunes shaped like a sickle. The open side of the sickle points in the direction in which the dune is moving. When the wind blows constantly from the same direction, it moves the masses of sand on the flanks of each dune faster than the sand in the center. Also, the velocity of the wind increases along the outer sides of the dune, and this accentuates the sickle shape. Barchans often travel across the desert in lines more than 60 miles long. Depending on the strength of the wind, they move from 65 to 165 feet each year. In their migrations, barchans can even overcome sizable obstacles. If these moving dunes encounter a high, steep cliff at the edge of a plateau, they at first come to a halt. Thousands of barchans pile up on top of each other, like breaking waves. Eventually they form a massive sandy ramp that gradually mounts to the top of the cliff. The endless chain of barchans can

march up a slope of as much as 25 degrees, and on their self-made ramp they will at last reach the high plateau, where they will continue their migration. They break up only when they reach an existing area of large sand dunes.

Another type of Sahara dune, the whaleback, is produced at an angle to the direction of the prevailing wind. These great cylindrical dunes can grow to a length of nearly 100 miles (see Plate 2).

Dunes also occur as a succession of ridges running parallel to each other and often stretching on for almost 200 miles. Between the dunes small strips of the rocky substratum can be seen. These corridors in the dunes are called gassis, and they run parallel to the direction of the wind. Such formations owe their existence to the trade winds that blow from northeast to southwest across large reaches of the Sahara, and consequently the chains of dunes all run in the same direction. The trade winds blow the gassis clear, while sudden gusts blowing crosswise tend to fill them with sand. When seen from the air, the longitudinal dunes appear to cross each other, looking like pigtails of sand that the wind has braided.

Oddly enough, of all the varied landscapes of the desert the sand wastes have been the least studied. There are many dune forms whose causes remain mysterious to this day; they certainly cannot have resulted from the effect of horizontal air currents. Try, for example, to build a pyramid of Ping-Pong balls six feet high. It is impossible. Yet the 1,000-foot-high star dunes are something of that sort. They likewise are built up of balls—myriads of tiny spheres of sand. Although no fully satisfactory explanation of their shape has been found, it is obvious that steady updrafts and electric charges between grains of sand play some significant part in their construction.

Similarly, the trade winds alone do not fully explain the longitudinal dunes that extend 200 miles. It may well be that the tremendous centrifugal force produced by the rotation of the earth, which sets the air and water of our planet in motion, also affects the movement and formation of dunes.

Another strange feature of the sandy wastes that to this day has not been fully explained is an auditory phenomenon—a rarely heard booming. It has been described by the British geologist R. A. Bagnold, who encountered it while he was studying dunes in southwestern Egypt, 300 miles from the nearest habitation. "On two occasions it happened on a still night, suddenly—a vibrant booming so loud that I had to shout to be heard by my companion. Soon other sources, set going by the disturbance, joined their music to the first, with so close a note that a slow beat was clearly recognized. This weird chorus went on for more than five minutes continuously before silence returned and the ground ceased to tremble."[*] Bagnold was never able to find a satisfactory explanation for the phenomenon.

According to my own estimates, between 3 million and 5 million tons of weathering material in the form of dust and sand are constantly in the atmosphere over the Sahara. This figure is, of course, only an average. It is quite likely that at certain times there is considerably less or more dust and sand in the air. About 80 per cent of the sand, which is heavier, accumulates within the Sahara itself, in the form of sand dunes; but the remaining 20 per cent—and, more important, most of the dust—leaves the Sahara region. I estimate that every day a million metric tons of dust and sand are blown out of the Sahara. If a freight train were to transport such a quantity, it would have to be a few hundred miles long. The greater part of the dust and sand is carried out over the sea and falls to the ocean floor.

But some dust from the Sahara drops back to earth, and this has significantly affected the melting and retreat of the alpine glaciers. The red dust that falls on the glaciers is deposited in a thin layer over the ice, reducing the reflection of the sun's rays so that more heat is absorbed by the ice. The glacier melts. Another influence of Sahara dust has been observed as far away as the Caucasus. Months after a severe sand and dust storm in the eastern Sahara, botanists discovered among the native plants of the Caucasus several species that normally grow only in the Sahara. Pollen and tiny seeds of Saharan plants had been transported along with the sand and dust—an interesting mode of plant dissemination. Desert storms often cause particles from the Sahara to land in unexpected places. Dust from the Sahara has regularly been noted in southern Sweden, and in 1970 one mighty Saharan dust storm actually crossed the Atlantic. For days, air traffic to several islands off the coast of South America was hampered by the Saharan dust.

The effect of the "sandblasting machine" within the desert itself is to expose deeper and deeper layers of the earth's crust. In our moist

[*] R. A. Bagnold, *The Physics of Blown Sand and Desert Dunes*, Methuen, London, 1941, p. 250.

1.  Sand dunes in the Sahara

climatic zone, vegetation and a thick layer of humus protect the firm bedrock of the crust. Only in quarries and where the earth's crust has folded into mountains can we see the rock of which it is composed. That is not so of deserts. There, almost the entire crust has been laid bare by the processes of weathering. In the course of millions of years, fantastic, bizarre landscapes of rock have been produced. An especially good example of the way crustal rocks are exposed and destroyed by weathering is to be found on the Tassili sandstone plateau in the central Sahara. This plateau, 300 miles long, 125 miles wide, and reaching as high as 6,500 feet, exhibits many rock formations that are typical of those found in deserts throughout the world. The word *plateau* is actually most inappropriate if we take it to mean a tablelike and generally smooth surface. When the area of the Tassili plateau was uplifted by movements of the earth's crust millions of years ago, it cracked into numerous blocks separated by deep, narrow clefts. The traveler finds himself in a bizarre landscape that has not its like on this earth. The sand-laden desert wind has transformed the plateau into a vast labyrinth of rocks in which a stranger would be lost forever without a local guide. There are several ravines as much as 2,000 feet deep, but the larger part of the area is broken by narrow canyons between 65 and 200 feet in depth; differential weathering has caused their formation out of the sandstone, the abrasive action of the wind modifying their shapes. The canyons run through the plateau like the streets of a large city, with "apartment houses" between them. The urban image is furthered by the fact that the rocks are riddled with a dense network of cracks that give the impression of masonry. There are intersections and blind alleys, and even bridges of rock crossing the "streets." Some of the lanes between the steeply rising walls of the cliffs are so narrow that a person can barely squeeze through. Other parts of the plateau are marked by caves and corridors cut by the "sandblaster." Just as in a city you can enter a large building from one street, walk through it, and come out on an entirely different street, so you can with the "buildings" of the Tassili plateau. By passing through corridors several hundred feet long, sometimes walking upright and sometimes crawling, you reach the next "street."

Sometimes several dozen narrow, deep canyons run parallel, separated by slabs of rock only a few feet wide but more than 150 feet

2. Cylindrical dunes in the Sahara

high. The wind has bored numerous holes in these walls. At other spots in the rock labyrinth the traveler finds himself entering a broad plaza where needles of rock 100 feet high stand scattered over the open space. Still other areas look like forests of giant mushrooms 10 or 15 feet tall. Like the needles, the mushrooms are produced by the abrasive action of the sand as it is whirled around by the wind. The largest grains of sand, which have the greatest abrasive surface, are heavier and are not carried as high as the smaller grains; this makes the abrasive action stronger down below, and the lower portions of the pillars are worn away faster than the upper portions. The result is the mushroom shape.

The Tassili plateau offers a view of every type of rock form the imagination can conceive (see Plate 41). To the traveler on foot, there seems no system at all to the endless variety of shapes in this rocky labyrinth. But from the air it can clearly be seen that the majority of the "streets" and longitudinal rock formations run in the direction of the trade winds, from northeast to southwest. And the surface of the plateau has an aerodynamic profile shaped by the wind. Its rock formations frequently resemble an enormous school of whales swimming with their backs out of the water. At its southern extremity the Tassili plateau gradually loses its special character. Only a few towers of rock, some of them as much as 500 feet high, have continued to resist the forces of weathering. These towers, characteristic of many deserts, are called zeugen, a German word meaning "witnesses," because they bear witness to the former extent of a plateau or stratum of rock. Wherever the zeugen, or "witness" mounts have been sandblasted away by the desert wind, the last remnants of rock quite often are found to have taken the form of glorious mosaic floors that stretch on to the horizon (see Plate 5). The sand-laden wind wears away any fragments of rock lying on the ground until they are level with all the rest and fit precisely into existing gaps in the substratum. When there are no more gaps, the remaining fragments are abraded entirely to sand. The individual stones of the mosaic pavement take every imaginable form; some are polygonal, with three, four, or six sides, some rhomboid, some round. Since the mosaic is made up of many different kinds of rock, it has patches where every stone is a different color. The even surface of the

3. Ahaggar Mountains in the central Sahara. The crags are the uneroded plugs of large volcanic necks, as the vents of volcanoes are called.

mosaic floor, which the drifting sand often polishes to a mirror finish, no longer offers any elevations that the wind can attack. These pavements of stone, which cover great areas of the Sahara and other deserts, represent the last stage in the weathering process. They are the thin base that remains of the once mighty strata of rock that have decayed to sand and dust.

## The Calendar of the Earth's Crust

On the flanks of weathering mountains, in the gorges of fissured plateaus, and along the cliffsides of zeugen, there is plain evidence that the earth's outer crust is built up of successive layers of various kinds of rock. These strata, formed by solidified deposits of such materials as mud and sand, have piled up over the course of the earth's long history. Every phase of the earth's evolution has left its traces, sometimes in the form of fossilized plants and animals.

In zones of greater rainfall, vegetation largely conceals these traces; in the desert they are clearly on display. The chronology of the planet has been exposed with extraordinary impressiveness in one of the most remarkable desert regions on earth, the weatherworn, fissured Colorado Plateau of Arizona. Here the agent of change was not the sand-laden desert wind but the waters of the Colorado River, which have been responsible for removing 14,000 feet of sedimentary rock. The result is a series of rocky landscapes whose wildness and beauty are unlike anything else on earth. The Colorado River, 1,500 miles long, rises amid the snows of the Rocky Mountains and for 1,200 miles flows through the arid sandstone plateau, before reaching the sea in the Gulf of California. The Colorado is the one great river other than the Nile that flows largely through a vast, coherent region of desert.

On its way through the plateau of red sandstone the river has dug a canyon of varying depths in the sedimentary rock of the earth's crust. The canyon was created over the millions of years during which the whole plateau was involved in the geological process that gradually lifted the Rocky Mountains and the sierras of the Pacific coast to their present maximum height of almost 14,000 feet.

4. Flat gravel desert, or serir, in the Sahara. Here the forces of erosion have destroyed everything; over millions of years every element that normally makes up a landscape has been leveled off.

The Colorado and its numerous tributaries in their old meandering beds counterbalanced the lifting of the crust by working their way deeper and deeper into the sandstone plateau. Meanwhile, the rising mountain ranges that now surround the Colorado Plateau prevented rain clouds from the Pacific from reaching the interior of the continent. That and a general thickening of the crust transformed the Colorado Plateau into what it is today: an arid relief desert.

In the northeastern part especially, the Colorado and its tributaries carved vast runnels into the sandstone plateau; this, along with the sandblasting of the desert wind, created a mighty stratified landscape containing a number of terracelike plateaus strewn with zeugen (see Plate 46). In the southeastern part of the plateau, the Colorado has eaten its way almost 6,500 feet down into the earth's crust for a distance of 250 miles, creating the Grand Canyon—the deepest erosion furrow on the surface of our planet. With its dizzying depth of a mile, its width of almost 20 miles, and cliff walls that often plunge more than 3,000 feet straight down into the abyss, it looks like a great gaping wound in the earth's crust.

The dome-shaped surface of the Colorado Plateau above the Grand Canyon represents a kind of oasis within an otherwise desertlike expanse. Its altitude is 8,000 feet, and during the harsh winters a large amount of precipitation falls in the form of snow. When the snow melts in the spring, the tough coniferous vegetation receives enough water to survive. The interior of the Grand Canyon, on the other hand, is absolutely arid. Because of the arched form of the plateau, the canyon receives little rain, and solar radiation, very intense during the cloudless period from spring to fall, heats the air between the canyon walls. The hot air within the canyon immediately evaporates the few drops of rain that fall. While the temperature on the plateau may drop to −40° C. in winter, the temperature in the interior of the canyon can rise to 50° C. (120° F.) in summer.

The formation of the canyon has brought about enormous climatic differences. If we include the 11,500-foot-high volcanic massif in the immediate vicinity of the canyon, the cleft from the peak of the volcanic cone to the lowest reaches of the canyon includes all the climates that would be encountered from the icy tundra zones of northern Alaska to the hot desert zones of Mexico.

Anyone who has stood on the brink of the Grand Canyon will find it hard to grasp that the tremendous abyss below is the work of a desert river. Yet this is so. Cloudbursts, rare but devastating, and desert winds erode the rocks of the Colorado Plateau, which are largely unprotected by any cover of vegetation, and carry the products of the weathering into the Colorado River. The dust, sand, and rock fragments, even huge boulders, turn the swiftly flowing Colorado into a reddish, whirling stone drill that eats into the earth's crust at varying speeds, depending on the composition of the rock. One of the hardest of rocks, granite, is worn away at the rate of just over 3 feet every 5,000 years. Day after day the Colorado transports 500,000 tons of sand and pebbles into the Pacific Ocean. A freight train with the same capacity would have to be 200 miles long.*

The Colorado has worn its way through all the strata of rock that have been deposited on the substratum of the continent in the course of the earth's long history. In the Grand Canyon itself, the river has already bored 1,600 feet into the substratum, which here consists of hard granite. The rocky walls of the Grand Canyon, those enormous stratified steps that lead like a giant staircase to the higher regions of the Colorado Plateau, provide a unique cross section of geological history. Over a billion years of the earth's history are revealed in the exposed strata of rock on the Colorado Plateau.

Within its strata of rock the earth has furnished us with a kind of diary whose entries cover the whole evolution of the planet. These "entries" consist of petrified plants and animals, which, along with the structure and composition of the various rocks, permit us to read the climatic and even the atmospheric conditions under which the mud and sand solidified into stone.

We cannot derive our chronology of the earth's history from the layering of rocks and sediments alone, since strata shift and are repeatedly worn away. For a true chronology, we need evidence of events that have happened only once and are not repeatable. The directional evolution of living forms provides us with just such events.

Early in the nineteenth century, the English surveyor, engineer, and canal builder William Smith discovered that similar strata of rock, though widely dispersed, occupying different positions, and of different thicknesses, could be identified by the presence in them of the same petrified life forms, or fossils. The evolution of life was not yet understood; but Smith saw that these index, or guide, fossils could be used to identify strata even if they had been displaced or overlaid by other strata. Thus a rough time scale of the earth's history

* At present the movement of sediment is interrupted by several dams. Instead of being washed into the Pacific, the sediment is temporarily stored in their reservoirs, which, as a result, will soon become silted up.

**Age of the strata in millions of years**

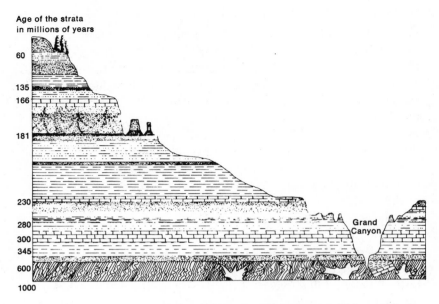

Cross section of the strata composing the upper part of the earth's crust in the Colorado Plateau

could be established. For his achievement in formulating the principles of biochronology, William Smith became known as the father of English geology.

One of the assumptions of biochronology was that the index fossils characteristic of a given geological period were distributed through strata of the same age everywhere in the world. The chronology could be regarded as even more definitely established if the evolutionary ancestors or descendants of the index fossils were found to be present in successive strata, so that an evolutionary series could be recognized. Unfortunately, such ideal conditions seldom occur.

The method of determining chronology by stratification permits the establishment of only a relative time scale. Not until the discovery of radioactivity, at the beginning of the twentieth century, was an absolute scale made possible. But that requires some explanation.

The atomic nuclei of certain radioactive elements decay into new elements independently of such external influences as pressure, temperature, and chemical changes. The element that begins the series is called the parent element; the products of decay are known as daughter elements. The rate of decay differs for each radioactive element and is measured by the time it takes for half the element to decay. This is called its halflife. The daughter elements may in turn decay into other daughter elements. Radium, for example, is one of the many daughter elements of uranium, a link in a well-known succession that ultimately ends in stable, nonradioactive lead. Uranium, with an atomic weight of 238, has the very long halflife of 4.5 billion years. In that time 500 of every 1,000 atoms of uranium will disintegrate into lead, with an atomic weight of 206. After another 4.5 billion years, the remaining uranium will again have diminished by half, and only 250 of the 1,000 atoms will be left. The number of lead atoms, however, will have risen to 750. Other elements have much shorter halflives, ranging from millions or thousands of years down to days or even seconds.

Physicists are able to measure the halflives with their highly sensitive instruments, and so the radioactive-decay series give us absolute measurements of the age of rocks. Elements with long halflives are used to date very ancient, deep rocks; those with shorter halflives serve for the dating of more recent rocks. In a mineral sample containing uranium, say, the daughter element necessarily becomes more and more enriched as time passes, while the parent element, the uranium, diminishes correspondingly. When we know the rate of decay, or halflife, and have a good estimate of the initial quantity,

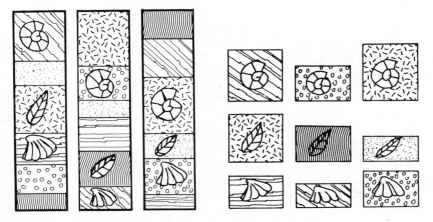

Model of biochronology after William Smith
(*Left*) Three cores of rock drilled at different places. The three different fossils found in each of them are embedded in different types of rock and in strata of different thicknesses. (*Right*) The fossils in the cores have been rearranged in their evolutionary succession, without regard to the type of rock or the thickness of the deposits.

the ratio of radioactive element to its decay product allows us to calculate the age of the mineral. The less uranium and the more lead the mineral contains, the older it must be. This way of calculating the age of rocks can, of course, be used only if the radioactive elements were locked in the crystals of the rock at the time of the original deposit. Most sedimentary rocks cannot be dated directly. The radioactive granules that were present at the time the sediments were deposited would already have had an indeterminate history of decay, since they usually come from older rocks that have been broken down by weathering, and therefore they cannot be dated as though they were "fresh" and "new."

That is not true of volcanic rocks, which are formed out of volcanic flux and usually contain traces of uranium. As soon as they cool below the melting point, all exchanges of material with their surroundings cease. The uranium and its decay products remain firmly bound within the solidified cold rock. From that moment on, the uranium clock begins to tick, like a stopwatch that has been set in motion by the pressure of a finger.

Rocks that contain the mineral glauconite can also be directly dated, since glauconite is newly formed during deposition. Age determinations for other sedimentary rocks, and for the fossil plants and animals embedded in them, can be made indirectly if they are found stratified between volcanic rocks or rocks containing glauconite. The drawing makes this clear.

The radiocarbon method, which permits the direct dating of organic substances, such as bones or wood, need not be discussed here, since the mere 50,000 years involved are of minor importance in the grand sweep of geological time spans.

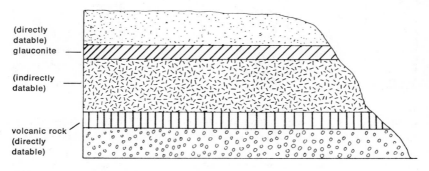

(directly
datable)
glauconite

(indirectly
datable)

volcanic rock
(directly
datable)

**Examples of datable strata**

Our present chronological techniques make it possible to determine, directly or indirectly, the real age of many rocks and fossils.
The ancient dream of natural scientists, especially of geologists and
paleontologists, has been fulfilled: when they dig up a saurian, they
can now say how many millions of years ago the animal walked the
earth. When they find a fossil in coal, they can tell when that particular seam was laid down.

In 1650, Archbishop James Usher reckoned that the creation of the
world had taken place at the stroke of nine o'clock on Sunday morning, October 23, in the year 4004 B.C. The scientists of the nineteenth
century, a good deal more realistic, estimated the age of the earth, on
the basis of the fossil record, as about 100 million years. Modern
researchers have assigned an age of more than 4 billion years to some
samples of rock.

For the scientist who knows how to read them, the strata of the
earth's crust and the fossils locked within them represent an inexhaustible potential source of information. In deserts, the forces of
weathering have exposed hundreds of layers, among them deep-lying
and very ancient ones that date back to the early days of the earth's
evolution. Normally, such strata lie many thousands of feet beneath
the surface. Thus the desert affords special opportunities for learning
about the evolution of our planet. But before we can understand the
earth's past and future we must take up the question of how this
planet, its atmosphere, and the life upon it came into being in the first
place.

# 2 / The Origins of Earth and of Life

The origin of the earth can be understood only if we know something about the history of the entire solar system. Yet to this day there is no theory of the origin of our planetary system on which all scientists agree. One reason is that astronomers and physicists have assembled such a multitude of details about the nature of the solar system that it is extremely difficult to find a hypothesis into which all the facts will fit. Some scientists are highly pessimistic about the chances of ever solving the exceedingly complex problem. The American astronomer G. P. Kuiper, for one, has pointed out that every theory of the origin of the solar system so far devised has contained assumptions that are very difficult, if not altogether impossible, to prove so many billions of years after the cosmic event.

Nevertheless, in spite of the enormous complexity of the physical relationships, I consider it quite likely that present efforts in space research, backed by close cooperation among the various scientific disciplines and by the application of sophisticated computers, will permit the testing of all the useful hypotheses that have been propounded so far. I am inclined to believe that we will see a definite solution to the problem in the near future.

With the aid of modern radio astronomy, for example, it has proved possible to look into the past of an important part of our solar system, the sun. Although this science cannot yet tell us whether other suns among the millions within our galaxy have planets orbiting them, it has allowed us to study the origin and death of such suns, or stars. (It is highly improbable, incidentally, that our sun alone has planets.) Our first impression is that the origin of our solar system can be described as a relatively simple physical process.

A striking fact about the nine planets of the sun is that all follow elliptical orbits around it and that these orbits all lie in approximately the same plane. That is a phenomenon permitting only a single conclusion: the sun and the planets were formed simultaneously, out of a cloud of cosmic gas and dust. Otherwise, according to the laws of celestial mechanics, the planets would circle individually around the sun, in different directions and along different planes, like moths fluttering around a source of light, each in its own orbit.

The cloud of gas and dust, which must have consisted largely of the primal substance of the universe, hydrogen, contracted steadily toward a common center under a force that is present throughout the cosmos, the force of gravitation. The continuous process of condensation raised the temperature of the huge masses of gas by millions of degrees.

According to another law of physics, a spherical cloud of gas contracting into what is called a protosun rotates faster and faster. Ultimately, the rotation of the tremendously hot sphere becomes so great that it flattens out into a gigantic rotating disk, which hurls a number of enormous fragments of gas away from its outer rim. These fragments, the substance of future planets, would have—in the far distant past—continued to move in the same direction and in approximately the same plane around the common center of gravity. As the principal mass of the disk of gas became stabilized again and continued to contract, gradually condensing into what is now our sun, the fragments also obeyed the law of gravitation and contracted toward their own centers of gravity. Such, presumably, was the origin of our planets. As plausible as this cosmological theory sounds, however, there are many physical phenomena in our planetary system that seem to stand in irreconcilable contradiction to it.

The most serious objection is the so-called paradox of rotational momentum, a concept that appears complicated but is relatively simple. Our sun today does not rotate as one rigid body; its parts rotate at rates varying from twenty-five to twenty-seven days. The protosun, however, must have rotated at a considerably greater speed in order to gain enough centrifugal force to hurl away the material that later formed the planets. The central mass of gas, the protosun, continued to condense after expelling the planetary masses. In fact, it must have shrunk steadily until it reached its present dimensions. At the same time its rotational velocity would have increased. By now, if our plausible-sounding hypothesis is correct, the angular momentum of the sun should be many times greater than it is—at least as great as that of all the planets combined. However, just the opposite is the case.

Although more than 99 per cent of the entire mass of the solar system is concentrated in the sun, the sun has only 2 per cent of the system's angular momentum. An interesting hypothesis propounded by Fred Hoyle and Hannes Alfvén in the 1950s may offer an explanation for this phenomenon, which otherwise appears to be incompatible with the laws of celestial mechanics.

The Hoyle-Alfvén theory is exceedingly complicated; what follows is a highly simplified summary of it. The two astrophysicists argue that the tremendous angular momentum of the protosun was transmitted in the form of magnetic lines of force across the enormous flattened disk to the fragments on the outer ring that were later to become the planets. As long as the center of the contracting gas cloud—what was to become the sun—rotated as fast as the disk and the separating portions, the lines of force were directed backward.

Most of the neighboring stars in the Milky Way system rotate as slowly as our sun, and it is possible that they, too, have lost their original angular momentum through the process of planet formation. If that hypothesis is correct in principle and can be confirmed by further studies, perhaps it can be proved indirectly, by measurements of the rotation of distant stars, that there are not only billions of other suns but also billions of other planetary systems in our universe, even though those planets, because of their small size, cannot be seen over the enormous distances.

There are other difficulties in the way of an acceptable explanation for the origin of the solar system. The sun's equatorial plane is inclined at an angle of about 6 degrees to the approximately common plane of the planetary orbits. This casts doubt on the assumption that the planetary masses were simply hurled away from the mass of the protosun by centrifugal force. Moreover, the chemical composition of the earth suggests that the surface of our planet has never, in its entire history, been hotter than a few thousand degrees centigrade, and therefore it could not have emerged from a cloud of gas and dust that was probably heated to more than a million degrees. Aside from that, so hot a gas cloud would be wholly unstable in space. Before it could cool and contract into planets, its components would dissipate at random into empty space. Only the mighty mass of gas gradually contracting into a protosun in the center of the hot cloud would be capable, because of its enormous gravity, of remaining coherent at such a high temperature.

It was this problem that led to the rejection of a hypothesis developed at the beginning of this century by the English astronomer James Jeans. Known as the "catastrophe theory," it occupied the minds of scientists for many years. Jeans was concerned with finding an explanation for the inclination of the orbital planes of the planets to the equatorial plane of the sun and for the superfluous angular momentum of the planets. He suggested that our planetary system had originated by pure chance, as the result of a cosmic accident: the original mass of all the planets was pulled away from the glowing surface of the sun by the gravitational attraction of another star, which had wandered so close to our sun that they had almost collided.

Jeans's theory seemed highly plausible for the simple manner in which it accounted for the surplus angular momentum of the planets. A force coming from outside might indeed have pulled the original masses of the planets out of the sun and thrown them into orbits in the same plane and the same direction. Moreover, the rotation that the planets would have gained appeared sufficient to account for their surplus angular momentum. Jeans's theory also served well as an explanation for the inclination of the planetary orbits to the equatorial plane of the sun. But later mathematical calculations undermined his theory in all respects. Furthermore, hot masses of gas torn from the surface of the sun in such a manner would not have been stable enough to form new planets. The gas would have expanded and dissolved instead.

Today, more and more confirmation is being found for a hypothesis whose basic features were developed not by an astronomer, but by the German philosopher Immanuel Kant, in 1755. He was the first to realize that the unitary nature of our planetary system and the unidirectional movements within it could be explained only in "historical" terms. Kant believed that at some time in the past all the elements of the solar system had been in physical contact. The German physicist and philosopher Carl Friedrich von Weizsäcker, by subjecting this hypothesis to calculations made possible by modern knowledge, has given it new prestige. The fundamental assumption, based on the physical facts described in this chapter, is that the sun and planets formed simultaneously but independently and that the planets did not originate from a hot mass of gas. Rather, an enormous

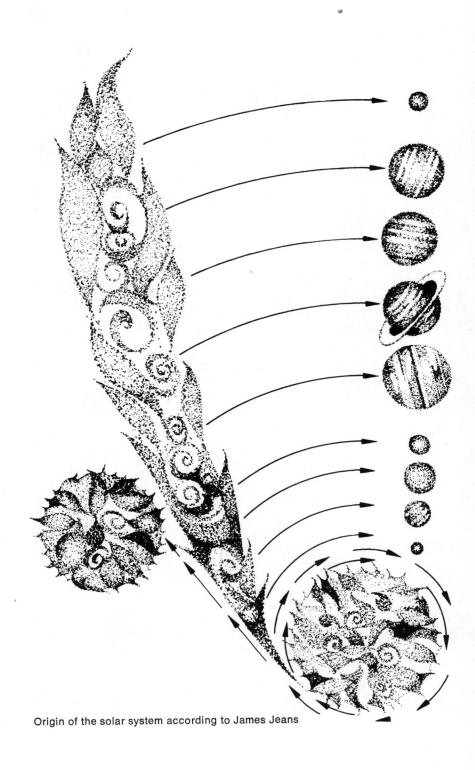

Origin of the solar system according to James Jeans

cold cosmic cloud, which consisted chiefly of hydrogen gas mixed with mineral and metal dust containing the ninety-two known elements, began rotating about a common center and flattening out steadily. A gigantic vortex formed in the cloud's center of rotation; numerous smaller vortices likewise formed in its outer reaches. Under the force of gravitation, the material in the vortices gradually became concentrated. The more massive each vortex became, the stronger grew its gravitational attraction, so that it drew to itself other parts of the cloud. In this way, the initially chaotic cosmic cloud developed a single great mass at its center, which would become the sun, and small centers of mass rotating around the common center, from which the planets would develop.

Again, however, an apparently insuperable physical obstacle arises. The most common element in the original cloud, hydrogen, could have developed sufficient gravity to form a stable body in the central vortex. But hydrogen is very easily dissipated; it seems doubtful that there would have been enough of this lightest of elements in the smaller planetary vortices for planets to condense. Our perplexity

Origin of the solar system by the formation of vortices

grows when we are reminded that the planets—at any rate, the inner planets of Mercury, Venus, Earth, and Mars—do not consist principally of the element hydrogen. The earth, the one we know best, of course, is made up chiefly of heavier elements, of silicates—compounds of silicon and oxygen—and of aluminum, magnesium, and iron. Only the central star of our system, the sun, is constituted largely of the light elements of hydrogen and helium; the heavy elements amount to only a hundredth part of its mass. When we take into account that stars, composed chiefly of vast masses of hydrogen gas, are far and away the most common heavenly bodies in the universe—there are more than 100 billion stars in just our own galaxy, the Milky Way—the tiny planets would seem to constitute a chemical anomaly in the universe as regards the distribution of their elements. They appear to be a concentration of heavy elements, tiny quantities of which presumably were distributed as fine dust in an original cloud consisting largely of hydrogen. But how could all the members of our solar system, the sun and the planets, have originated from a single cosmic cloud and yet have ended up with the elements distributed so differently among them? That poses a major difficulty for the theory that the sun and planets originated from a single cloud.

Nevertheless, an explanation is possible, in terms of the history of the system. Within the planetary vortices that gravity gradually transformed into rotating spheres of gas, a sorting out of the elements by weight must have taken place. Let us follow this further course of evolution on one of the inner planets, our own earth.

The traces of the heavy elements, in the form of mineral and metal particles, sank toward the center of the sphere, forming a small, massive core surrounded by the light gases hydrogen and helium. This sphere of gas, the protoearth, had a diameter 2,000 times larger than that of the present earth, and the sphere consisted chiefly of the light gases.

At the same time, the great vortex of gas in the center of the entire cloud—the protosun—was contracting steadily and also assuming the shape of a rotating sphere. Remember, too, that in this central sphere of gas more than 99 per cent of the total mass of the system was—and still is—concentrated.

As the result of gravitational force, a pressure of 200 billion tons and a temperature of 15 million degrees centigrade built up in the interior of the gaseous sphere. A vast nuclear reactor came into being—a sun. The inconceivably high pressure and temperature initiated a process of nuclear fusion in the interior; hydrogen was burned as nuclear fuel and energy was released.

Up to this point, the development of the planetary system had taken place in almost complete darkness, illumined only faintly by the light of distant stars. Now the sun began to shine, and the atomic fire in the center of the system showered the whirling masses of the relatively cool gas clouds—the newborn protoplanets—with light and warmth. The new sun would henceforth govern the further evolution of the forming planets and would link all the members of the solar system indissolubly together, on into the remotest future.

The effect of proton radiation from the sun—the "solar wind"—on the gaseous sheaths of light hydrogen and helium around the protoplanets caused the breakup of the layers of gas, and their components scattered into space. What remained were the small planetary cores consisting of heavier elements, in the form of fragments of minerals and metals.

The inner protoplanets, those closer to the sun, received the greater part of the stream of solar energy; the outer planets were bathed in a mild, weak light, all that reached them from the distant sun. The giant planet Jupiter still has a chemical composition that, astronomers believe, corresponds to the character of the protoearth 5 billion years ago. The huge body of this planet consists mainly of hydrogen and helium, probably surrounding a small core of heavy elements. It is believed that the composition of Jupiter has hardly changed since it condensed out of the cosmic cloud of gas and dust. Since Jupiter is more than 425 million miles from the sun, the solar energy impinging upon it has always been too weak to accelerate the hydrogen molecules of its gaseous envelope enough so that they could escape the planet's mighty gravitational field.

The inner planets would never have grown beyond the dimensions of small fragments of rock had not another process begun simultaneously with the dissipation of their envelopes of light elementary gases. The planetary cores that remained had sufficient gravity to attract much of the mineral and metal dust—consisting of a wide variety of elements—left over from the cosmic cloud. More and more particles of dust flew toward the cores and adhered. The original cloud probably contained, in addition to the elementary gases, a large number of sizable fragments of rock and metal, in the form of meteorites ranging in size from pinheads up to diameters of several miles.

Most of the dust and mineral fragments that rained down on the protoearth consisted of silicates and mixtures of iron and nickel.

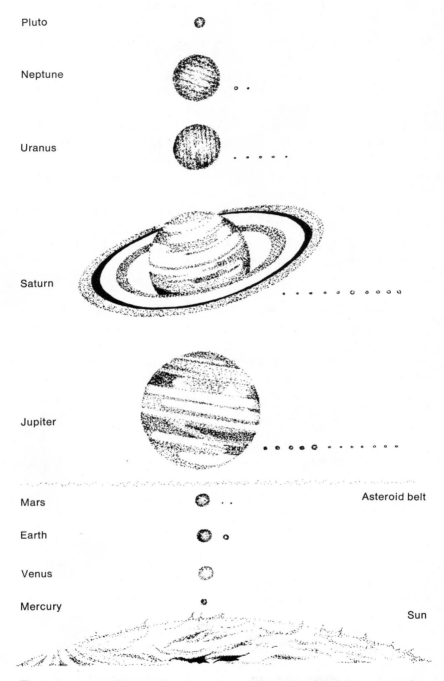

Pluto

Neptune

Uranus

Saturn

Jupiter

Mars                                     Asteroid belt

Earth

Venus

Mercury

                                          Sun

The solar system. The drawing conveys a rough notion of the sizes of the nine planets and their moons in relation to each other and to the central star of the system, the sun.

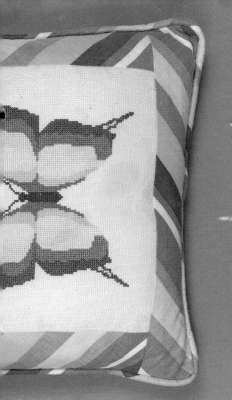

**Green and Blue Butterfly**
8 × 10

**43. Japanese Butterfly**
4 × 4

**45. Double Hsi**
3½ × 3½

**10. Hsi Symbol and Butterfly**
12 × 12

Ibex designs P.O. Box 556, Southport, CT 06490

Small quantities of other elements, some of them radioactive—uranium and thorium, for example—were also present. In addition to the long-lived radioactive elements, which take billions of years to disintegrate and therefore are still present, there were many short-lived radioactive elements, which soon began decaying into the more common nonradioactive elements. The considerable amounts of energy this released, along with the energy of gravitational contraction and that from the constant impact of meteorites, brought the steadily growing earth to the melting point. The melting of minerals and metals resulted in still another sorting out of the elements by weight. The iron and nickel melted sooner than the silicates and, because of their higher specific gravity, gradually sank to the center of the protoearth, while the light, rock-forming silicates rose to the surface. From them the thin crust and continents of the earth would be formed.

For millions of years our earth in its primal state must have been a terrifying place. Large and small meteorites came crashing onto the viscous, red-hot, glowing surface of the protoearth, splashing fiery masses of matter high into what was not yet the air. Lighter, cooler, and harder slabs of rock built up of silicates floated like black ice floes on a red ocean in the midst of the inferno.

On earth, the complete melting meant that the features characteristic of meteorite bombardment were erased. But many scientists believe that traces of that phase can still be seen on the heavenly body closest to ours, the moon. Unlike the earth, the moon probably never melted totally because of its smaller mass. Since, for the same reason, it also was never able to form and hold an atmosphere, it has had nothing like wind and weather to reshape its surface, as these forces have constantly reshaped the surface of the earth. That is why traces of its origin, the scars of meteorite impact, have remained preserved on the moon's surface for billions of years in the form of innumerable craters.

Radioactive determinations of the age of moon rocks have shown that the moon had a solidified crust some 3.5 to 5 billion years ago, before the meteorite bombardments ended. The large number of craters preserved in the protocrust of the moon is excellent evidence that small cosmic bodies were significantly more numerous during the creation of the solar system than they are now, and that not only the mass of the moon but also that of our earth could have been formed out of dust particles and mineral fragments rushing together under the force of gravitational attraction.

By the time the supply of capturable particles was used up, our planet had reached its present size. Its surface cooled slowly and, like the surface of slowly cooling milk, formed a thin skin. The earth's skin—its crust—is made up of floating rock, and is quite stable for all its thinness. Datings of the oldest earthly rocks indicate that the crust had attained a high degree of solidity as early as 4 billion years ago.

As is evident to anyone who has seen a small meteor flare and burn out in the atmosphere, the earth is to this day collecting particles of unknown origin—the last sparse remnants of the original cosmic cloud from which it sprang. The total mass of the meteor shower that falls on the earth annually nevertheless amounts to about a million tons. Now and then, at long intervals, larger fragments hundreds of feet in diameter and weighing tens of thousands of tons fall to earth and gouge enormous craters in the crust. One example is the Diablo Crater in the Arizona desert, which is almost a mile across and over 600 feet deep.

It was only a chain of fortunate accidents at the time of our planet's origin that kept it from being condemned, as the moon was, to remain without atmosphere, without oceans, and without life.

As we shall see, the earth came into being in a particularly favored place in the solar system, where it would receive the "right" amount of solar radiation for its further development. At the time of its origin it also acquired exactly the "right" mass, which, in combination with its favorable distance from the sun, was to provide the basic requirement for all life: the formation of oceans and an atmosphere.

## The Genesis of the Oceans and the Atmosphere

Let us imagine the earth reduced to the size of a peach. Without too great a violation of scale, the seed within the peach pit would correspond in size to the earth's inner core, which is probably solid iron and nickel. The thick hard shell of the pit would correspond to the outer core, which probably contains liquid iron. The meat of the peach is the viscous rock of the earth's huge mantle, which is made up of a great variety of minerals. The thin outer skin of the fruit, however, far exceeds in thickness the earth's solidified crust, which is, at most, about 40 miles thick. The satiny fuzz on the skin brings us back to scale again, for it corresponds to the thickness of the earth's atmosphere. On the other hand, the oceans, with their average depths of only 10,000 feet, would hardly dent the skin of the peach. If we painted them on in blue watercolors, the thickness of the layer of paint would exceed by many times the relative thickness of the earth's layer of water, no matter how thinly we brushed on the color.

All life, all the things we can see and that constitute our experience, takes place on the surface of the thin, solidified crust—on the continents, in the oceans, or in the atmosphere. Without the atmosphere there would be neither wind nor weather, and no circulatory system to pump water, in the form of rain clouds, out of the oceans and transport it to the dry continents. Without water in liquid form, stored in oceans, ponds, and rivers and in all living substance, life would be inconceivable. Water is the universal solvent in which the chemical processes that sustain life go forward. Without the carbon dioxide of the atmosphere, plants would not be able, with the aid of sunlight, to produce the food that is the basis of existence for all living things. Without the water vapor and oxygen of the atmosphere, which act like great invisible filters in keeping the sun's ultraviolet radiation away from the earth's surface, no highly evolved life forms could exist on our planet.

How did these essential and potentially evanescent gases come into being and remain attached to the planet?

When the formation of the globe was largely completed and solar radiation had swept away virtually all the hydrogen and helium that had formed the first atmosphere, there remained a slowly cooling and solidifying crust that, as the Book of Genesis describes it, was waste and void.

The earth's formation of a second atmosphere—hereafter referred to as the protoatmosphere—was possible only because the fugitive light elements tend to combine chemically with heavier elements. This meant that the rock of the crust and the mantle was able to hold large volumes of the lighter gaseous elements, preventing them from dissipating into space. Had it been otherwise, there would be not a drop of water on earth. For water is nothing but the chemical combination of the lightest element, hydrogen (H), with the heavier but also gaseous element oxygen (O). And carbon dioxide, a gas in our atmosphere that is one of the compounds most essential to life, is a combination of oxygen with carbon (C).

There is one interesting exception to the tendency of the gaseous elements to combine easily. The so-called noble gases, xenon, krypton, argon, and neon, do not combine chemically with other elements. It is highly indicative of the processes that took place during the formation of the earth that the noble gases are extremely rare. They are underrepresented in the mix. The reason is that they were not able to "save" themselves by combining with other elements, as large quantities of hydrogen did, when the earth lost its first atmosphere through solar bombardment and the melting caused by radioactivity and the heat of meteorite impact. The relative rarity of the noble gases is a further link in the chain of evidence that the earth originated in the manner I have outlined. Calculations show that the water content, and thus the supply of elementary hydrogen, still bound in the rock of the earth's crust and mantle amounts to $4 \times 10^{18}$ metric tons, or three times as much water as is in all the oceans. It was from the enormous reservoir of light elements stored in its interior that the earth, in the course of its evolution, formed its second atmosphere.

The interior of the earth has remained a glowing mass of partially molten rock to this day, causing enormous pressures to build up under the thin solidified crust. The high pressures and temperatures beneath the crust have created gigantic "safety valves" in the form of volcanoes. Without volcanoes there would be no oceans on earth, no atmosphere, and no life. Not only do volcanoes bring streams of fluid rock to the earth's surface; also, as can be seen by the clouds of smoke thousands of feet high above an active volcano, their lava, once it is relieved of the great pressures on it, releases vast amounts of water vapor, carbon dioxide, nitrogen, and several other gaseous elements. During the earth's early period, when the heat from the millions of meteorites had not yet escaped into space and the supply of "fresh"—not disintegrated—radioactive elements was vastly greater than today, there must have been far more volcanic activity. Through innumerable volcanic chimneys and cracks in the slowly cooling crust, many of the light elements bound in the rock of crust and mantle reached the surface in the form of gaseous compounds. Then as now, from 80 to 90 per cent of the volcanic exhalations probably consisted of water vapor. The remaining 20 per cent or so would have been chiefly carbon dioxide, nitrogen, methane (a gaseous compound of carbon and hydrogen), and ammonia (a compound of nitrogen and hydrogen), as well as remnants of free hydrogen.

At the time the protoearth was venting its protoatmosphere, this must have been a truly hellish planet. Since the surface was probably cooling very slowly, the water vapor that condensed at great heights was never able to reach the surface as rain. Much like the ghost rains in the desert, the raindrops would evaporate when they struck the layer of hot gases immediately above the crust. As the volcanic chimneys continued to supply more water vapor, over the course of millions of years larger and larger amounts of seething, swirling vapor accumulated in the protoatmosphere. Soon a cloud of steam wrapped the entire planet, preventing the sun's rays from penetrating to the surface. For hundreds of thousands of years, the gloomy, dark wastelands of this vacant protoearth would be illuminated only by brilliant flashes of lightning during the incessant thunderstorms.

But the surface continued to cool, and at last rain succeeded in reaching the ground. What followed was a series of cloudbursts that, for perhaps several million years, flooded the steaming crust of the earth with water (see Plate 8). The water collected in great hollows in the crust. Such was the origin of the oceans. Along with the rain, many other chemical components of the atmosphere entered the newly formed oceans. The dense layer of water vapor continued to descend; individual clouds formed for the first time, the sky became visible, and the rays of the sun illuminated the surface of the earth.

It was in this phase of the earth's evolution that its distance from the sun proved of optimal value. Had our planet chanced to come into being somewhat closer to the sun, the solar radiation would have been considerably greater and would have prevented the condensation of the atmospheric water vapor. At a greater distance, the water vapor would have become so cool that it would have created a thick armor of ice around the entire globe. In either case, the most vital medium of life, liquid water, would have been lacking.

Yet for all that conditions in this early stage of the protoearth's evolution were relatively favorable to life, neither human beings nor any of the plant and animal life we know today could have survived a minute in the atmosphere as it was then. We would instantly have suffocated, and the biological building blocks of which every part of our bodies is composed would quickly have been destroyed by the ultraviolet component of the solar radiation, which reached the surface of the earth almost unhindered. One gas was missing among all the gases pouring out of the volcanoes and constituting that early atmosphere. It was the one elementary gas without which no highly evolved life on our planet would have been possible—the gas that sustains our breathing and fends off the deadly ultraviolet radiation: oxygen.

The importance of oxygen for breathing is self-evident. Its role as a filter for ultraviolet radiation is also apparent to anyone who has done mountain climbing. At several thousand feet above sea level, the oxygen filter between living flesh and the sun becomes a good deal thinner. A frequent consequence is severe skin burns, unless protective ointments are used to make up for the lack of oxygen cover.

The absence of free oxygen from the earth's protoatmosphere at first seems something of a mystery, in view of the fact that the element oxygen constitutes about 46 per cent by weight of the shell of rock, or lithosphere. That is a vast supply of oxygen. The silicates, which constitute 92 per cent of the minerals in the crust, consist largely of spherical packages of oxygen ions. But the oxygen is too firmly bound in the silicate structures for it to be liberated under the conditions of pressure and temperature that prevail today. Breaking down the silicates is possible only under very high temperatures—probably higher than those present in the lithosphere of the proto-earth. It also requires the presence of a strong reducing agent, such as carbon. And even then free oxygen would not be released; the oxygen would come out of the rocks combined with the reducing agent, carbon, in the form of carbon monoxide or carbon dioxide. Clearly, the free atmospheric oxygen that is indispensable to our metabolic functions could not have reached the atmosphere of the earth as a component of the volcanic gases.

The fact is that for several billion years the protoatmosphere contained hardly any oxygen. That has been confirmed by studies of some very old iron-ore strata, which were laid down several billion years ago and until recently were buried under other sediments, so that they were never in contact with the present atmosphere of the earth. From the worn, rounded forms of the ferriferous rocks, it is obvious that in the early days of the earth they lay on the surface and were exposed to the weathering influences of the atmosphere as it then existed. But the rocks do not show the degree of chemical change that would have been inevitable if they had been in contact with an oxygen-rich atmosphere. In the latter case, the iron would have combined with the atmospheric oxygen, would immediately have been oxidized, since oxygen is highly active chemically. The absence of such chemical activity in the ancient iron-bearing rocks proves that our earth did not initially have an atmosphere rich in oxygen.

We now know that our oxygen-rich atmosphere represents an exceptional stage of planetary evolution. Free atmospheric oxygen was not essential for the next decisive step in the history of the earth, the development of life. On the contrary, because of its chemical activity, its ability to combine rapidly with so many other elements, oxygen is a dangerous poison, and it would have been a fatal threat to the first simply constructed building blocks of life. Modern scientists agree that if ample oxygen had been present in the earth's protoatmosphere from the beginning, life would never have come into being.

Free oxygen in the amounts we find today appeared in the atmosphere relatively late in the earth's history. And the amazing thing about this gas is that it was first produced and released by the early life forms themselves. It was a waste product of their metabolism, quite worthless—in fact, dangerous to life because of its toxic effect. The early life forms would take several billion years before they learned how to deal with this environmental poison that they themselves were producing. Finally they succeeded in making their dangerous metabolic waste product useful to their further evolution by instituting a recycling process.

We human beings are the temporary end product of that recycling. But let us not anticipate. First we must ask: how did the earliest forms of life arise?

## The Origin and Evolution of Life

One reason scientists today know as much as they do about the origin of life on our planet is that the decisive step toward the evolution of life in effect took place twice: once in the remote past, about 4.5 billion years ago, and once in 1953, in Chicago, Illinois.

Generations of scientists had speculated on the origin of life. Recognizing as they did the extraordinary complexity of the natural laws that underlie the functioning of all living organisms, scientists despaired, as recently as the middle of our century, of ever being able to account for the origin of life billions of years ago. At such a remove in time, there could be little evidence left in the form of mineral deposits or microfossils.

Some sort of scientific voyage back to this distant past became possible only when, during this century, physicists and biologists began unraveling the universal laws that, independent of time, govern the functioning of the smallest units of matter and of life. That accomplishment is one of the great intellectual feats of the human race. With the knowledge gained, it at last became possible to lift the secret of the origin of life out of the darkness of the past and make it accessible to modern scientific research.

First, a word of caution. The problem of the origin of life can by no means be regarded as solved in all its details. But scientists have on the whole come to agree on the basic features. Here we will sketch the exceedingly complicated subject only as it relates to our own purpose: a better understanding of earth's future fate as the desert planet.

Stanley Miller, while still a student of biochemistry in Chicago in the early fifties, had the good fortune to make one of the century's crucial scientific discoveries. In the course of his studies, Miller had frequently encountered the multitude of confusing and complex hypotheses about the origin of life. All of them eventually came up against one seemingly insoluble problem: the chemistry of life is extremely intricate, and only living organisms themselves appear able to synthesize the vital building blocks known collectively as biopolymers. Yet those vital compounds would have had to exist before life could begin. How, then, could such compounds have arisen, at the beginning of the earth's evolution, out of the lifeless chemical compounds in the protoatmosphere and the proto-oceans of earth, that is, out of water, carbon dioxide, methane, and ammonia?

Miller, after pondering these questions day and night, decided to undertake a basically simple experiment that somehow had been overlooked by the specialists who had been studying the problem for decades and had become obsessed by its complexities. Miller took a glass retort and filled it with methane, ammonia, and water, that is, a miniature version of the probable protoatmosphere of the earth. He hermetically sealed the retort, to keep the contents from any contact with the present atmosphere. Now, in order for the chemical components in the retort to react, as chemists say, a source of energy was needed. The protoatmosphere of the earth had been influenced by two principal sources of energy: ultraviolet light, which is the part of sunlight richest in energy, and the electric discharges of the thunderstorms.

Since it does not matter in principle what source of energy is chosen for a chemical experiment, Miller decided in favor of lightning, which he sent into his miniature atmosphere in the form of high-voltage electric discharges.

Eager to find out what chemical reactions might have taken place in his protoatmosphere, he switched off his apparatus after a single

5. Mosaiclike stone pavement of the desert

6. In the Sahara's "ocean of sand"

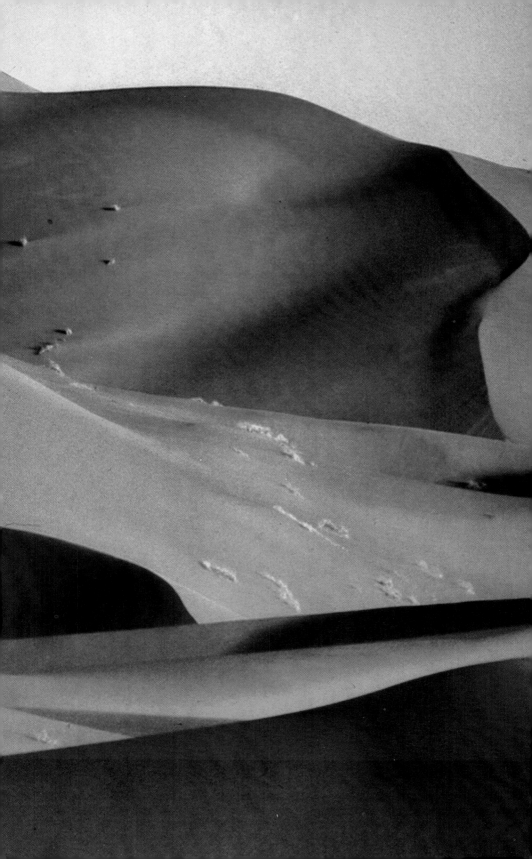

7. The desert of the Grand Canyon in Arizona shows plainly how the earth's outer crust is built up of many layers of rock that were deposited one on top of another over geological ages.

8. Singer terraces in the volcanically active Yellowstone Plateau. Without the dead trees in the foreground this might be from the early ages of earth, when enormous quantities of water vapor were exuding from the interior and gradually condensing, until the first rains fell on the planet's steaming surface.

9. Hot springs and geysers on the Yellowstone Plateau

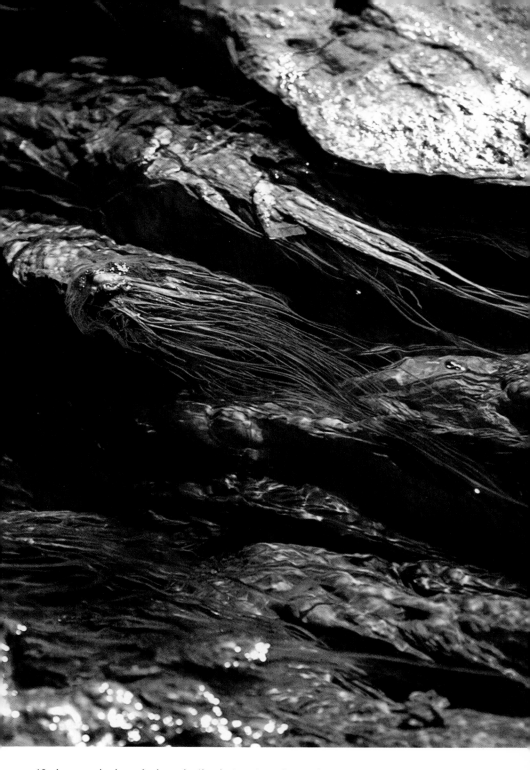

10. Long colonies of algae in the hot water of a spring on the Yellowstone Plateau

day, opened the container, and began a chemical analysis of the contents. The result was a scientific sensation; the news spread almost instantaneously among the world's biochemists. During the experiment inorganic molecules of methane, ammonia, and water had given rise to large organic molecules, the amino acids glycine, alanine, and aspragine. The amino acids—there are some twenty of them—are the fundamental building blocks of life. The universal living substance known as protein is made of them.

Stanley Miller, in his simple experiment, which he had probably begun without especially high hopes, had virtually re-created the process of primal creation and the inception of our own existence at least 4.5 billion years ago on this earth.

Initially, of course, many of Miller's colleagues questioned the student's technique and undertook to duplicate the experiment. They achieved similar positive results. The significant factor was the employment of all the elements of which the earth's protoatmosphere had been composed. When several scientists, among them such well-known biochemists as Wilhelm Groth, Hanns von Weyssenhof, and A. N. Terenin, treated their retorts with ultraviolet light instead of Miller's electrical discharges, amino acids were repeatedly produced.

It seems paradoxical that, of all sources of energy, deadly ultraviolet radiation should have produced the building blocks of life out of inorganic substances. But the danger from ultraviolet radiation lies in its ability to break up molecular combinations and reduce the molecules to their constituent elements. The elements, however, have a tendency to recombine. When the inorganic molecules were broken up by ultraviolet light during the experiments, natural laws caused some of them to recombine, not into their starting compounds of water, methane, and ammonia, but into the larger molecules of amino acids, the building blocks of life.

Now let us return to that period on earth of 4.5 billion years ago. The breakup of inorganic molecules by ultraviolet light and their subsequent recombination into amino acids was, of course, a two-edged sword. If those tiny building blocks of life were not to be

11. The petrified shell of a goniatite, proof that regions of the Sahara were flooded by the sea in an earlier geological age

12. Fossilized cone-shaped coral reef and stratified marine deposits in the northwestern Sahara. To appreciate the scale of the picture, note the person standing in the lower left corner, where the dark brown marine deposits abut the light-colored coral reef.

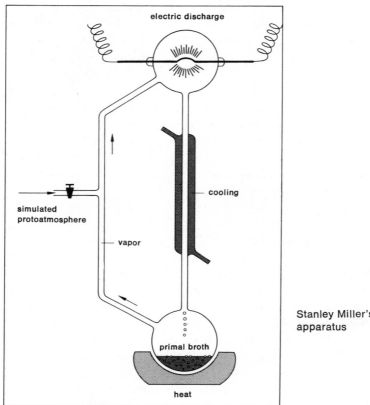

electric discharge

cooling

simulated
protoatmosphere

vapor

Stanley Miller's
apparatus

primal broth

heat

broken down again into their initial components by the same energetic radiation, they somehow had to be withdrawn as quickly as
possible from the influence of the ultraviolet rays. How might that
have been accomplished?

Remember that free oxygen in its present atmospheric concentration was not present to act as a shield against ultraviolet light. In any
case, an atmosphere that was rich in oxygen from the start would
have hampered the nonbiological origination of those building blocks
of life, and if they had come into being by some other means the
oxygen would have destroyed them promptly through its chemical
activity. More paradox.

At this point, it must again be stressed that the relationships are
extremely complicated so that it is difficult indeed to describe the biochemical interactions taking place on earth. The difficulty is that
innumerable natural processes are going on simultaneously and that

all of them are inextricably intertwined. Anyone who has really thought the problem through must basically agree that what we mean by nature cannot be described by us who are a part of the whole. It is like trying to describe the entire biological nexus of a tree whose leaves are constantly changing their position in the summer breeze and among whose branches birds and swarms of insects are flying. It is impossible to describe the state of this biosphere and the relationships among all its components, the externally visible and the internally invisible biochemical processes, at one given point in time, let us say at 4:00 P.M. on July 4, 1976. Before the first word of the report could be written, countless changes would have taken place.

As a first step, we may assume that the amino acids that formed in the atmosphere fell, enclosed in raindrops, into the ponds, lakes, and oceans that were beginning to fill the hollows in the earth's surface. In this phase of the earth's evolution, that is, the building blocks of life literally rained from heaven.

But some nonbiological production of amino acids would also have taken place in the top layers of the growing bodies of water. The ultraviolet light would penetrate the water to a depth of several feet, and the necessary basic elements were present not only in the atmosphere but also, in solution, in the water. In the course of time, sizable amounts of amino acids accumulated in the waters of the earth, until eventually they were as rich in organic substances as a broth. But, as in the atmosphere, wherever amino acids were formed the same energy that had brought them forth broke them up again. Water, however, because of its fluid nature, was a much more propitious environment for repeated recombination than the atmosphere was. That may be why individual large molecules of amino acids began to react with each other. They attached themselves together into larger molecular chains, the biopolymers, out of which proteins would arise.

Extensive laboratory experiments have proved that such a picture is not mere speculation. If the initial organic substances of life are exposed for a sufficiently long time to the energy of ultraviolet light, more and more complex organic molecules arise. Within a short time after Miller's first experiment, scientists had artificially created a large number of the organic compounds that are known to be components of presently existing organisms. Among the complicated molecular chains were, for example, adenosine triphosphate—chemists call it ATP for short—which stores energy in the cells for the crucial life processes, and even nucleides, those elementary components of the nucleic acids that serve as the basis for heredity. Another molecule created in the biochemists' retorts was one that almost certainly was present in the waters of the earth more than 4 billion years ago, porphyrin. Porphyrin molecules would become a component of the green chlorophyll of plants, furnishing the basis for our own existence.

This combination into larger and more complicated compounds, which biochemists call polymerization, must be viewed as essentially no more than a quantitative effect. The decisive qualitative leap—primal creation—had been taken when the amino acids were formed by nonbiological processes.

Now let us return to the question we asked earlier: how were some of the biopolymers able to save themselves from the destructive energy that had just brought them forth? That they were not broken up almost as soon as they were made was probably due to currents that carried them from the dangerous upper strata of water into deeper levels. At a depth of around 30 feet the ultraviolet radiation would no longer have had any effect, since that depth of water offers much the same degree of protective filtration as our present oxygen-laden atmosphere. Presumably the water at that depth became an ecological niche for the biopolymers. Here, in the course of ages, they could develop undisturbed into the first primitive one-celled organisms.

When those first organisms came into being, they were at once confronted with the problem of food. They probably solved it in simple fashion by feeding on the substance of the primal broth in which they were floating and from which they had arisen. That is, they devoured the amino acids that ultraviolet light was still producing in the uppermost layers of water and that the currents carried down to them. Their primitive metabolic function consisted in breaking the newly created amino-acid molecules down again and using the molecules' binding energy as their own vital force.

## The Development of a Modern Atmosphere

The nonbiological production of amino acids was accompanied by an entirely different process in the earth's protoatmosphere, one that was also to be of decisive importance. It was an effect of ultraviolet light called photodissociation, and it resulted in the inorganic production of the first free atmospheric oxygen. When this energetic part of the sun's radiation broke up atmospheric gases, the subsequent recombinations only rarely yielded amino acids. By far the greatest share of the atmospheric components, especially the water-vapor molecules, took a different chemical route. When molecules of water vapor were split by ultraviolet light, the released hydrogen dissipated rapidly into space, while the heavier component, the oxygen, remained behind in the atmosphere.

It was always possible, of course, for the components of the water molecule to recombine. But, because of the tendency of the light element hydrogen to leak off into space, the amount of oxygen in the atmosphere was bound to increase. Since free atmospheric oxygen is an extremely effective ultraviolet filter, however, it screened off the very energy that was producing it and prevented the decomposition of more water molecules. And so the production of oxygen in this manner soon came to a standstill.

Moreover, the oxygen could not enjoy its freedom for long, because it is so active chemically. Metals made up a large part of the earth's crust, and when the oxygen came into contact with the barren surface of the protocontinents it immediately oxidized the metals there—that is, it chemically combined with them.

Although by far the greater part of the heavy metals had sunk to the center during the formation and melting of the earth, huge quantities of metal, chiefly iron, remained distributed among the rocks of the crust. Had it not been for that, we would have no iron ores to exploit today. The enormous store of metal of which the earth's core is composed will surely remain unavailable forever.

The oxidation of iron consumed much of the free atmospheric oxygen. Ultraviolet light could again penetrate the atmosphere unhindered, and the process of oxygen production by the decomposition of water molecules began anew, followed again by the oxidation cycle. The cyclical increase and cessation of oxygen production amounted to self-regulation of the oxygen content of the protoatmosphere. However, this cycle could never have led to the present level of oxygen in the atmosphere. Precise calculations have shown that such inorganic production of oxygen would have kept the atmosphere's oxygen level fluctuating at around a thousandth of what it is today. That insignificant supply would not have sufficed to oxidize the rocks of those ancient iron-ore beds. It should be added that this primitive form of oxygen production still takes place in the highest strata of the atmosphere.

The appearance of free oxygen in the atmosphere initially made the earth less, rather than more, favorable to life. To be sure, the oxygen acted as a shield against ultraviolet light and thus was beneficent. And in preventing further decomposition of water-vapor molecules the oxygen acted to preserve the hydrosphere (all the water vapor in the atmosphere plus all the liquid water on earth). Water provided a suitable environment for the primitive organisms and was, of course, the prerequisite for all further evolution of life on our planet. Nevertheless, the level of oxygen in the air of the protoearth (0.1 per cent of the present value) was not a sufficient shield, and the oxygen itself was a danger to primitive life.

One might think that the organisms would have been safe from the deadly radiation no matter where they were in the water, whether at a great depth or on the surface. But that was not so. The reason is that ultraviolet radiation is only a part of the broad spectrum of electromagnetic radiation emitted by the sun. The wavelengths to which we are sensitive, visible light, are another part of that spectrum. The wavelengths radiated by the sun range from radio waves many miles long to gamma rays, whose waves are shorter than a billionth of a centimeter. Physicists measure wavelengths in what they call angstrom units; one angstrom unit equals a ten-millionth of a millimeter. The area within the total spectrum of electromagnetic radiation that the human eye sees as visible light extends from 4,000 angstrom units, which we see as violet, to about 7,000 angstrom units, which we can barely perceive as dark red. As the word *ultraviolet* implies, shortwave ultraviolet light is just beyond the region of violet that we can see; it extends from 4,000 down to 100 angstrom units. Within this range is the special radiation responsible for the splitting of water molecules and the inorganic production of oxygen—dissociative wavelengths, they are called—which lies between 1,500 and 2,100 angstroms.

Wavelengths above that range continued to pass unhindered through the thin oxygen shield of the atmosphere. Calculations show that, given the ancient atmosphere's oxygen level of 0.1 per cent of the present level, ultraviolet radiation above about 2,000 angstroms would have had little difficulty penetrating to the earth's surface. And those are precisely the wavelengths to which all organisms are most sensitive. Adequate protection against ultraviolet radiation was to require a rise in the oxygen content of the atmosphere, as well as the formation of ozone—oxygen molecules containing three atoms of oxygen.

In this early state of the evolution of our earth, such protection would not have offered the primitive organisms any significant advantages—would not, say, have allowed them to rise to the surface of the water, let alone venture out onto the ·dry, desertlike protocontinents. To have done so would have put them in deadly peril from the chemical activity of the new free atmospheric oxygen. Even as it was, there was considerable danger. A dense stratum of water shielded them from ultraviolet rays, but it could not serve as an absolute barrier to the atmospheric oxygen, because oxygen is soluble in water (if it were not, fish could not breathe). The reason these early living organisms were not killed off by oxygen in the water probably is that the greater part of the oxygen was chemically bound by the iron in the earth's crust.

Oxygen that did penetrate into the sphere of the living most likely was captured and rendered harmless by the vital inorganic basic substances that had participated in the creation of life: the methane and ammonia dissolved in the water. These compounds lose their stability in the presence of free oxygen. Oxygen quickly combines with them, and the chemical reactions leave carbon dioxide and nitrogen dissolved in the water. This buffer effect of methane and ammonia protected the first primitive organisms against direct contact with free oxygen.

Nevertheless, amino acids and biopolymers were probably "burned up" by free oxygen more rapidly than they could be resupplied through the action of the volcanoes. What saved the first organisms from death by oxygen also meant the end of the nonbiological creation of life. From that point on, life could arise only from other life.

The stages described in these few pages did not, of course, all occur in swift succession, but were taking place simultaneously over inconceivably long periods of time, surely hundreds of millions of years.

The primitive organisms had barely escaped perishing by oxygen death when they were exposed to a new and no less dangerous test of their ability to survive. Because the newly created oxygen was consuming important components of the protoatmosphere and preventing the further production of amino acids by ultraviolet light, the organic components in the primeval broth also diminished steadily. As we have noted, the first primitive unicellular organisms fed on the amino acids, and the slow drying up of their source of nourishment must have constituted the equivalent of a world-wide famine. My being able to write this account and your being able to read it are proof that the primitive organisms did not succumb. We owe our existence to the fact that our remotest ancestors in the primeval waters found something akin to a raft equipped with everything they needed for survival. That "raft," in the figurative sense, was the porphyrin molecule mentioned earlier.

Scientists assume that the porphyrin molecule was from the beginning as common in the primeval waters of the earth as the other organic molecules of nonbiological origin. It is also assumed that some of the original cells used these molecules as building blocks. One type of cell continued to feed on the inorganically produced amino acids. But another type, the cells containing porphyrin, made perhaps the most important advance in the entire history of the earth: they developed photosynthesis. In the course of evolution, virtually all life on earth has come to owe its existence directly or indirectly to this biochemical process.

The cells containing porphyrin had the ability to absorb the visible portion of the sun's outpouring of energy. With the captured solar energy, each cell created a kind of chemical factory within itself and learned to make the organic nutriments that were fast disappearing from its environment. Out of simple inorganic molecules, the cells built up, with the aid of solar energy, more complex organic molecules. The nutriments they created were chiefly starches, fats, and proteins; the initial inorganic constituents were carbon and the chemical compounds of oxygen and hydrogen.

If we examine a carbohydrate, one of the end products of photosynthesis, we find that it is nothing but a compound of carbon, hydrogen, and oxygen: $C_6H_{12}O_6$. Carbon and oxygen were conveniently available in the carbon-dioxide molecule ($CO_2$), which was abundant in the earth's atmosphere and waters from the very beginning. The protoplants probably obtained their indispensable hydrogen from the energetic inorganic compounds that were also the source of the amino acids: methane, ammonia, and one we have not mentioned before, hydrogen sulfide. In addition to the hydrogen that the protoplant cells were so eager for, these compounds contain carbon, nitrogen, or sulfur, but no oxygen. That fact is highly important for a full understanding of evolution. When the protoplants extracted the hydrogen from these compounds there was no residue of free oxygen.

As the first universal food crisis mounted and conditions of life in the primeval waters steadily deteriorated, the hour of the protoplants came. Since these cells, each with its own food factory powered by solar energy, were independent of the ineffective and vanishing food supply produced with the aid of ultraviolet light, they were able to survive.

But if we are to understand the future of life on earth we must also consider the fate of the other type of primitive cell—the type that apparently was less successful in the struggle for survival and seemed in danger of extinction. At the risk of enormous simplification, that fate can be described in a phrase. These cells profited in the simplest fashion from the great leap forward taken by the other type of cell: they ate the protoplants. The first animal cells, the ancestors of all human beings, had come into existence.

## Plants from Volcanoes

Let us take a moment to see whether it is possible to determine when the first photosynthetic processes took place. Like most of the developments in the history of the earth, photosynthesis has left distinct traces in the rocky journal of the earth's crust. In certain parts of the crust that lie exposed in the deserts and steppes of South Africa, and that are known to be 3.2 billion years old, there are unmistakable signs of this key advance of the living organism. The traces in the rock—what are called chemofossils—consist of the basic molecular structure necessary for photosynthesis, porphyrin, and a typical vegetable food product, organic carbon.

Since we can safely assume that the rock in which such fossils are found was once a part of growing sediments accumulating over geological ages, the real beginning of photosynthesis must probably be dated still earlier, perhaps 3.6 billion years ago. Moreover, the process of photosynthesis is extremely complex biochemically—scientists are still puzzled by many of its aspects—and it must already have gone through a long period of development before its first demonstrable appearance 3.2 billion years ago and its presumed beginning 3.6 billion years ago.

Photosynthesis, which probably developed as described in the preceding section from the necessities of the first food crisis, may well have been one of the earliest "inventions." There may even have been some elementary stages of photosynthesis before the oceans were formed.

That hypothesis is rejected by most scientists, since all the earliest processes of life were dependent on the protection afforded by water. It is my opinion, however, that the indispensable water in liquid form may have been available elsewhere on earth before precipitation caused it to accumulate in the great basins of the crust. There must have been some places where the water vapor spewed out by the volcanoes condensed and gathered. But where could that have been, on a rocky crust hotter than 100 degrees centigrade, which turned raindrops to steam the instant they struck the surface? The only places possible were the craters at the top of the volcanic cones, thousands of feet above the hot lowlands, in the higher and some-what cooler regions of the atmosphere. It is quite conceivable that during the inactive phase of a volcano its high crater might have cooled sufficiently so that the water vapor still puffing out of the volcano could condense to form a crater lake. But even under the coolest conditions imaginable, the water in such a lake would have been close to boiling. Could that liquid but extremely hot environment have sheltered the early forms of life?

Before I go into that, let me note that the height of the volcanoes in this earliest period of our earth's history may never be known. Since volcanic cones are built up of loosely heaped lava and ashes, they are worn down rather quickly by the forces of weathering and, as a rule, do not persist through long geological ages. It is likewise not known whether the earth's crust between 4 and 4.5 billion years ago had become thick and stable enough to support volcanic cones thousands of feet high.

Still, there are factors that raise the possibility of life's having be-gun under such circumstances—raise it above the level of mere speculation. Some scientists hold that the various types of organisms in those hypothetical hot crater lakes have survived more than 4 billion years, down to the present, in the hot thermal waters of regions that are volcanically active today. Thermal waters are found in almost all the volcanic regions of the earth; they are common in New Zea-land, Iceland, Japan, Java, and Kamchatka. But nowhere is their concentration so impressive as on the Yellowstone Plateau, in the western United States. On this wooded plateau, 6,500 feet high and often shaken by earthquakes, the surface of the ground is separated from the earth's molten interior by only a relatively thin crust of rock. Among the dense evergreen forests lie hundreds of emerald, sap-phire, and turquoise lakes, clear as glass and boiling hot; above the tops of the tall firs, hot water and steam from geysers shoot higher than 300 feet in the air (see Plate 9).

Most of the seething, hissing water in such lakes and geysers is ordinary ground water that has trickled into cracks and underground hollows of the crust and been heated by the molten magma until it boils to the surface or, as in the geysers, bursts out explosively with the force of hot steam. But a certain amount of it is probably what geologists call juvenile water and is entering into circulation on the earth's surface for the first time. This water rises, in the form of steam, out of the molten magma of the interior, just as it did 4.5 billion years ago. Through cracks in the crust it reaches the deep chimneys and hollows that supply the geysers, and it emerges into them and into the boiling lakes along with the steam from the ordinary ground water. In addition to water, other compounds from the magma, such as the gaseous carbon dioxide, methane, ammonia, and hydrogen sulfide that once formed the protoatmosphere of our earth, pour out into the steaming lakes. The last three compounds do not remain stable long; they are soon consumed by the free oxygen that is now available in such great quantities. That is just one of the reasons why the existing thermal waters—in contrast to the hypothetical thermal waters of the early earth—cannot repeat the original creation of life.

But it is highly interesting that innumerable unicellular organisms—bacteria and blue-green algae—live in the hot chimneys of geysers, and especially in the boiling lakes. Scientists who specialize in the primitive life forms have in recent years found many indications that these microörganisms may be direct descendants of the life forms with which life began on the still hot protoearth. Among the indications are that most of the bacteria in the thermal waters can endure the temperature of boiling water and that some species carry on a primitive form of photosynthesis in which they extract their necessary hydrogen not from the water but from the hydrogen sulfide dissolved in it. And they manage to do this speedily, before the hydrogen sulfide is consumed by the free oxygen that is also in solution in the thermal waters.

The blue-green algae that live in thermal waters likewise carry on photosynthesis and can endure temperatures as high as 75° C., or 167° F. (see Plate 10). Such traits, some scientists conjecture, could well have been acquired on a hot protoearth. If these conjectures should be confirmed by further research, it would follow that the steps toward the development of life need not all have taken place in the early oceans. Indeed, those steps might have started much earlier, in the hot crater lakes, which provided all the requisite basic substances. Thousands of feet above the lowest portions of the crust, the dense cloud of hot water vapor that shrouded the entire globe may occasionally, in this early phase, have thinned enough to permit visible sunlight to reach the surface of the crater lakes, supplying the energy essential for the process of photosynthesis. From those crater lakes, the first vegetable and animal cells might have been carried by streams of water into the hollows of the crust and been washed into the newly forming oceans.

But whatever the locus of the origin of life, there is no question that its further development took place amid the safe expanses of the new oceans.

## The First Environmental Crisis

The first real photosynthesis to be carried out on an extensive scale, that of the blue-green algae, had dangerous as well as advantageous side effects. It will be recalled that the "modernization" of food production, as a consequence of the first world food crisis, consisted in plant cells' developing the capacity to extract hydrogen from the omnipresent water molecules by splitting them, using the energy of visible sunlight for the purpose. The cells could not use the oxygen released in this process, and they promptly vented it as a worthless by-product of their chemical factories. But, as we have seen, in the early phases of the development of life free oxygen was a deadly poison because of its intense chemical activity. The oxidizable bodies of the primitive life forms were in danger of burning up in the by-product of their own metabolism.

The risk was relatively low at first, when there were only a few blue-green algae exhaling oxygen in proportion to the immense supply of water in the new oceans. But as the algae, in their highly successful struggle for survival, increased enormously in number, greater and greater amounts of oxygen were released into the environment, until an environmental catastrophe fatal to all organisms, including the animal cells, seemed inescapable. Help was at hand, however, and for a while the oceans were able to act like gigantic sponges, absorbing and neutralizing the dangerous toxin.

By the time most of the vast quantities of water vapor vented by the volcanoes had cooled, condensed, and accumulated in the hollows of the earth's crust, largely completing the formation of the oceans, something like modern weather patterns had begun to develop on earth. A vast system of water circulation was initiated, with the sun acting as the pump. Its heat radiation evaporated water from the surface of the oceans and lifted it into the atmosphere in the form of water vapor. As the vapor cooled, rain clouds formed; driven by winds, they transported their water from the ocean basins onto the flat, desertlike continents.

No sooner had the thin crust of rock formed over the earth's molten interior than the forces of weathering began to wear down that rock wherever it protruded, in the form of level protocontinents, out of the water of the oceans. Wind and running water from the rains carried the products of weathering, in the form of sand and dust, to the oceans.

Much of the weathered rock consisted of iron in bivalent form— ferrous oxides that are not yet saturated with oxygen. As mentioned earlier, vast quantities of iron remained in the solidifying crust even after most of the iron sank to the core of the earth. When this iron was carried into the ocean waters and encountered the oxygen excreted by the blue-green algae, the iron was further oxidized, to the trivalent form, and became ferric oxide.

It will be recalled that the small amounts of oxygen formed inorganically in the atmosphere by the energy of ultraviolet light had been sufficient to convert only an infinitesimal part of the iron in the continental rocks to the trivalent form. That left plenty of ferrous oxide around, and it was this bivalent iron that, when it was washed into the oceans, made them into an "oxygen sponge." The ferrous oxide absorbed the environmental poison that the organisms were producing. That gave life the time—a great deal of time, as we now know—to grow habituated to the dangerous gas in its environment.

Ultimately, a point was reached in the earth's evolution when the myriads of blue-green algae were releasing so much oxygen into the oceans that the iron could not bind it chemically as fast as it was being produced. The excess oxygen was vented from the oceans and enriched the atmosphere. The slow but constant increase in the oxygen content of the atmosphere led to still further oxidation of the iron on the continents. If an observer had been able to watch the earth from the very beginning, he would have seen how this process gradually colored the surface of the desertlike continents a deep red. For the first time in the earth's history, the characteristic red coloration of the desert appeared.

An added consequence of the oxidation on the continents was a halt to the steady supplying of oxidizable iron to the oceans. This meant that the oxygen content of the water would begin to rise once more. And, inexorably, oxygen death for all the organisms living in the oceans would have followed—if they had not, after several billion years, learned to include their excretory product in a kind of recycling process that permitted better utilization of the food they produced for themselves by photosynthesis. That is, they had made their excretions useful to their further evolution. Both the plant and the animal organisms in the seas began breathing in the environmental poison that oxygen had once been. (At this early stage in the earth's story land organisms did not yet exist.)

Before we look into the far-reaching consequences of the development of breathing, let us see how the first plant cells and the first animal cells—which ate the former—had managed to derive their vital energy from carbohydrates without using free oxygen.

Scientists are able to explain the process with great precision because descendants of these primitive life forms, the anaerobic bacteria, still exist alongside the more evolved forms. The anaerobic bacteria obtain their energy not by breathing but by an archaic process of breaking down carbohydrates that biologists call fermentation. The cells start with a complex food molecule built up from simple substances during photosynthesis. In some ten additional chemical reactions, they decompose the molecule into its components, releasing chemical binding energy with each reaction.

But anaerobic bacteria carry the decomposition of carbohydrates only as far as an intermediate step, breaking them down into pyroacemic acid and two related compounds, lactic acid and alcohol. Without the employment of chemically active oxygen, the breakdown of food, and therefore the utilization of its chemical binding energy, ends here. Much of the energy goes unused.

In other words, the anaerobic bacteria have chosen a highly uneconomical method of breaking down the nutritive substances that result from photosynthesis. All other cells were sooner or later forced by the appearance of oxygen in their environment to continue the chemical activity, carrying the breakdown of the still energetic compounds through a further series of steps. At this period in the evolution of the earth, the cells had virtually no alternative, since vegetative photosynthesis was continuing to produce so much oxygen. If they were to avoid having the entire substance of their bodies, including their food, spontaneously oxidized in a single chemical reaction—in other words, burned up—they had to seize control of the dangerous chemical activity of oxygen and incorporate it into a series of reactions, which we call the respiratory cycle. In developing the respiratory cycle, a series of some thirty reaction stages, the plant and animal cells provided themselves with a controlled form of oxidation that could supply their own energy needs. In terms of survival, they made a virtue of necessity.

Breathing plant and animal cells continue the process from precisely the point at which the anaerobic bacteria stop their energy-liberating decomposition of the food molecule: with pyroacemic acid.

Even today, however—leaving aside the bacteria that are strictly anaerobic—every breathing cell, including the cells of our own bodies, continues to employ anaerobic fermentation as an archaic stage in its metabolism. If we equate the ten stages of the fermentation reaction with the lower stories of a building, we can say that the development of the earth's atmosphere has added only thirty more stories—the thirty stages of the respiratory reaction—to the original structure. Employing the great chemical activity of oxygen, breathing cells carry the breakdown, or burning, of carbohydrate molecules from the pyroacemic-acid stage on back to the simple compounds out of which the food molecule was originally built by photosynthesis: to carbon dioxide and water.

In other words, breathing is the exact reversal of photosynthesis. To sum up: from carbon dioxide and water molecules, plant cells, using the energy of visible sunlight, build up a large and complex food molecule. Since this molecule contains less oxygen than the initial compounds, some oxygen has been left over as a waste product. In a kind of recycling process, both plant and animal cells breathe in some of the oxygen and use it to burn up the food molecules that they either have produced for themselves or have ingested, at the same time releasing the binding energy of the molecules. The cycle is completed.

In contrast to the archaic cells that practice only fermentation, breathing cells, since they are able to exploit the entire binding energy of the food molecule, have at their disposal fourteen times as much energy for their vital functions. Naturally, that gave the earliest breathing cells a decisive advantage in survival over the fermenting cells. Their more effective metabolism was a step toward the higher evolution of life. Breathing made it possible to supply energy for a sizable association of cells, something metabolism by fermentation could never have accomplished.

So it was that at a particular point in the earth's development the rising oxygen content of the oceans provided the trigger for the further evolution of life forms toward their present elaborate organization—the provisional end point apparently being ourselves. Breathing probably developed about 600 or 700 million years ago, at the turning point from the Precambrian to the Cambrian period, when the oxygen content of the ocean waters and of the atmosphere rose to about 1 per cent of its present value: the level at which respiration can occur. The effect that the development of breathing had on the further differentiation of life can be read in the record of rocks from this geological period that have been exposed to the sandblasting of the desert wind. In such rocks a large number of extremely varied fossils of plant and animal life begin to appear. It happens so abruptly that the scientists have called the phenomenon the Cambrian explosion.

Photosynthesis was a kind of voluntary response by life to the first world food crisis. The resultant waste product, however, actually forced respiration on life, and in so doing forced its higher evolution. From then on, from the time that more abundant oxygen compelled life to adopt a higher level of energy consumption, life was literally dependent on free, unbound oxygen. But because of oxygen's unstable, unreliable character—to this day, the greater part of the free oxygen released by photosynthesis promptly combines chemically with the rock of the earth's crust—the more highly evolved, breathing life forms have fallen into a highly risky dependency. In geological terms, their very existence is endangered.

The evolution of successively higher and more differentiated life forms was, of course, dependent on an increasing supply of energy, which could be obtained only by the oxidation of the products of photosynthesis. But it is inherent in the laws of the universe that movement takes place toward the condition that is more probable. And the more probable condition is always the one that is more undifferentiated and chaotic. It is the condition that arises spontaneously, without any input of energy. Long-term preservation of the higher life forms by such a successful process as respiration seems to be is—on a geological time scale—extremely improbable. Someday in the future oxygen may be withdrawn from the organisms that have become wholly dependent on it. That day seems to be drawing closer with the steady expansion of the deserts, the causes of which we will be discussing.

For the time being, the shift to oxygen respiration that occurred some 600 million or 700 million years ago has brought this result: more and more human beings, because of their ever increasing demands for more and more energy, are directly or indirectly dependent on a constant increase in the supply of free oxygen. This stage in our geological evolution is most plainly expressed in the doctrine that continued industrial growth is an absolute necessity. But if, along with the inevitably increasing industrial consumption of oxygen, photosynthesis diminishes and less and less oxygen is released, the earth's fauna will not be supplied with all the oxygen they require. In fact, because of the spread of deserts and the rising industrial consumption, plants are releasing less and less oxygen. Moreover, the weathering of desert rocks has made the deserts large consumers of oxygen. But all this will be discussed in more detail in later chapters.

How do cells—the cells of our body, for example—behave when their highly evolved and sensitive respiratory mechanism is endangered by an unfavorable oxygen supply or by the unchecked intake of chemicals, either as components of breathed air, as drugs, or under the influence of stress? The sensitive respiratory mechanism suffers irreversible damage. The cells retreat, as it were, to the archaic phase of their metabolic functioning; they go back to fermentation. Apparently acting under external compulsion, cells with damaged respiration activate those ancient, outmoded metabolic functions dating back to the eons when the environment contained no free oxygen.

This spontaneous relapse of breathing cells to archaic metabolic habits describes the disease of cancer, the scourge of our century. All those who seek a steadily rising standard of living are participating in the steady decrease of free oxygen in our environment, and they must not be surprised if more and more cells abandon respiration and retreat to fermentation—become cancerous.

A well-known biochemist has suggested that cancer is not a genuine disease at all, but represents the fate of highly differentiated and highly sensitive breathing cells in a cosmic environment subject to constant pressures toward antidifferentiation and disorder.

In a sense, the desert represents an advanced stage of this cosmic impulse toward the more probable state: disorder.


## Coral Reefs in the Desert Sand

Suppose someone told you that in the heart of the planet's greatest desert, the Sahara, he had seen gigantic tropical coral reefs surrounded by fish and other marine animals. You would surely assume that such a story was the product of an overheated imagination. Yet the story is quite true.

In the northwestern Sahara is an extensive tropical coral reef, resembling an atoll, that rises some 300 feet above the sun-seared waste of stone and sand. It is a petrified coral reef that was built between 380 million and 400 million years ago, in the Devonian period. Its architects were myriads of tiny lime-excreting coral polyps living under the waters of a warm tropical sea.

Similarly, the surface of the rocky wasteland in which the reef rises is the petrified mud and sand bottom of the ancient ocean. That explains why the land in this part of the Sahara is literally strewn with the petrified remains of thousands of marine animals that once populated the ocean around the coral reefs. The fossils, especially the large, many-colored spiral windings of the goniatites' shells and the longitudinal shells of the orthoceratites, have been brought to a high polish by the sandblasting of the desert wind, making the region one of the most beautiful parts of the Sahara (see Plate 11).

This apparent paradox—that one of the driest regions of the earth lay beneath the sea in an earlier geological age—is by no means an exceptional circumstance on the planet earth. More than a third of the vast surface of the Sahara consists of petrified marine deposits from many different geological ages. That is true of all deserts and, in fact, of all other continental areas. For—the matter will be discussed in greater detail in the next chapter—the earth's crust is in constant movement. We regard the ground beneath us as a solid foundation for all our everyday activities. But the frequent reports of catastrophic earthquakes should serve to remind us that we are actually living on the surface of a thin rocky skin that covers a glowing hot sphere.

Over the course of geological time, forces originating in the earth's mantle have repeatedly broken the solidified crust into numerous huge plates. In the process, parts of continents have been lifted up and folded, or have temporarily sunk and been flooded by the sea for millions of years. In the most ancient periods of the earth's history such temporary floodings of great continental regions took place more easily, probably because the crust in the places where it formed the continents was significantly thinner than it is today.

Huge areas of the Sahara have sunk and been covered by the sea as many as eight times in the course of the earth's long geological history. The skeletons, shells, and armor of the organisms living in those oceans were deposited on the sea bottom, along with the sand and other sediments that the wind and rivers transported from the continents to the seas. Over the 50 million or 60 million years during which parts of the African continent were under water, deposits of stone hundreds of feet thick were built up from these sediments. Then the same motive forces that had produced the sinking caused the continental areas to rise gradually. As they rose, the petrified sediments often were tilted and displaced, and broke up into huge sedimentary blocks. When exposed to the forces of weathering on the surface of the desert, the marine deposits were worn down and carried away again, but in reverse order of their deposition—that is, from top to bottom. The most recent deposits, from the last 200 million or 300 million years, were carried away first, eventually exposing the coral reefs of the Devonian period, dating back 400 million years. In their vicinity some even older sediments have been exposed, marine deposits as much as 500 million years old from the Silurian and Ordovician periods. The coral reef in the Sahara consists of half a dozen isolated cones over 250 feet high that look like huge sugar loaves—they are called biohermae—and stand one behind the other in a line that stretches for miles.

Although exposed for millions of years to the destructive forces of weathering, the cones have retained their original form because they are made of an especially hard, crystalline limestone. Other marine sediments have been largely worn away. Moreover, in contrast to the loosely consolidated sedimentary rocks in the vicinity, which are composed of pebbles, sand, mud, and the remains of dead organisms, the coral polyps that built this reef gave it maximum solidity.

The marine sediments that were deposited nearby while the reef was growing are still there, in the form of huge, dark brown slabs of rock tilted at a sharp angle (see Plate 12). The beautiful calcareous shells of the goniatites and orthoceratites, the armored scales of primeval fish, the trilobites—reminiscent of primitive crabs but already having highly developed faceted eyes—and hundreds of other species of animals and plants have remained for 400 million years locked in the dark ocean sediments. These creatures, and the mighty coral reefs themselves, are impressive evidence of the great advances that the development of respiration had brought, even at this early period in the history of the earth.

Because of the excellent state of preservation of the desert coral reef, all the details of its various biospheres can be clearly seen. On the sea bottom, the trilobites scratched in the mud hunting for food. The scratches are preserved to this day in some of the slabs of rock. Above this level is the horn coral reef, named after the small individual corals, about 4 or 5 centimeters long, that lived here, close to what was formerly the sea bottom. Topping that is an extensive plat-

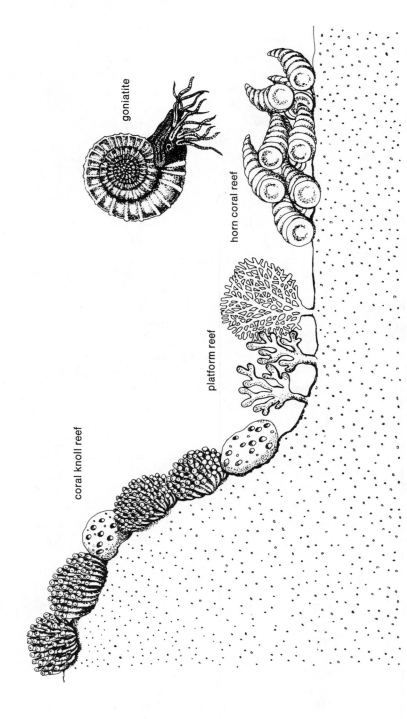

goniatite

horn coral reef

platform reef

coral knoll reef

Biosphere in the primal ocean, and the coral reef. Individual corals are shown larger than life-size in relation to the other life forms.

form reef, which consists of branching and fan-shaped corals. Many of the branches are now shattered and form large masses of stone rubble surrounding the tall, cone-shaped main reef. This so-called coral knoll reef is composed of huge, compact colonies of corals.

The horn coral reef, with its delicately curved, cornucopial corals, is especially revealing. Wherever these noncolonial corals lie, half embedded in stone—that is, in what once was the mud and sand of the sea floor—most of them are pointing in the same direction, which is the direction of the ocean current at the time they were alive. From the positions of these corals we can read, 400 million years later, information on the currents of the ancient oceans. Even more amazing is that each one of these small individual corals constitutes a calendar in which every day of all the years of its lifetime is recorded. We have been able to extract from those calendars a great deal of knowledge about the past of our earth, and about its future.

We owe this astonishing knowledge to the American geologist John W. Wells. Close study of living corals has shown that this organism, which builds its stony exoskeleton out of secretions of lime from its "foot," forms its shell in consonance with the rhythm of the seasons. The same was true of the individual corals that built up the Sahara reef of the Devonian period. The naked eye can discern on the outside of their shells many circular grooves and ridges, which correspond to the annual rings that are exposed when a tree trunk is sawed through. The number of annual rings is equal to the age of the coral animal.

When Wells examined the clearly defined annual rings under a microscope, he discovered that each was subdivided into diurnal sections, which formed because the coral polyp always ceased its excretion of lime at nightfall. When the scientist counted the diurnal

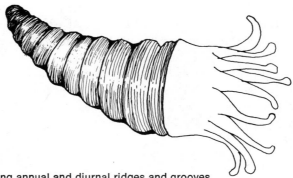

Coral animal showing annual and diurnal ridges and grooves

sections, he made a staggering discovery. Between 380 million and 400 million years ago, the year was 395 days long, not 365!

Since we assume that the earth's orbit around the sun has not changed significantly in the last 400 million years, and therefore the length of the year has remained the same, it follows that the day in the Devonian period must have been shorter, in order for 395 of them to fit into a year. Calculating the length of this day is a simple matter. If the year had 8,760 hours then, as it does now, dividing that number by 395 gives about 22 hours as the length of the day. In other words, the rotation of the earth around its axis must have been faster 400 million years ago than it is today. And, in fact, it has been shown by the methods of modern physics that the earth's rotation is slowly being braked over long geological ages, so that the day is growing longer. The causes of this slowdown and the dramatic consequences it may hold for the future of our earth, especially for the formation of deserts, will be discussed in the last chapter.

But shortly before the coral reefs formed in the area that is now the Sahara an event of tremendous importance took place.

## The Conquest of the Land

Some 410 million years ago, plants began to advance from the edges of the oceans onto the continents. That crucial event brought the end of the primeval desert, which until then had covered the land. For 4 billion years the continents had been a wasteland, without a trace of life, except perhaps in the crater lakes of the volcanoes. No land area on earth had ever known the cooling shade of a green tree, the dancing flight of a butterfly across a flowering meadow, the song of a bird. The land was a primitive waste bombarded by deadly ultraviolet rays, its only events the weathering caused by rain and sporadic storms of sand and dust. That is how it was for 4 billion years. Such an immense span of time is beyond the power of the human mind to imagine.

What made possible the conquest of the land?

The waters of the ancient oceans were flooded with sunlight, for the oceans were much shallower than they are today. They were filled with green algae, which over eons of photosynthesis released more and more oxygen, steadily enriching the atmosphere. The supply of oxidizable materials on land, such as bivalent iron, had by no means been completely consumed. But oxidation of the materials of the crust proceeded at a much slower pace on dry land than in the water. For a while the consumption of oxygen on the continents and its production in the oceans reached a state of equilibrium. And it was precisely this equilibrium that gave life its unique chance to move out onto dry land. By the Upper Silurian period, enough vegetative metabolic excretion had accumulated in the atmosphere for its oxygen content to reach about 10 per cent of the present level. That was enough to shield living cells from most of the deadly ultraviolet radiation.

This was probably the period of the first formation of a stratum of ozone in the upper atmosphere—without which no highly evolved life on land would have been possible. Ozone ($O_3$) is an active molecule of oxygen made up of three oxygen atoms, as against the two atoms in the normal molecular oxygen of the air ($O_2$). The extremely thin layer of ozone at an altitude of about 12 miles would be only a fourth of a millimeter thick if it were subjected to the pressure prevailing at the surface of the earth. Yet it is able to absorb the greater part of the deadly ultraviolet rays. The way ozone works is graphically depicted in the drawing on this page.

By now it is widely known that the fragile layer of ozone is threatened by the fluorocarbons that are used as propellants in spray cans.

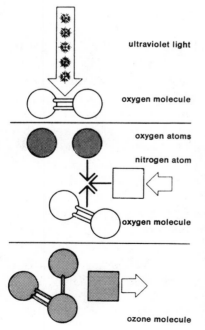

ultraviolet light

oxygen molecule

oxygen atoms

nitrogen atom

oxygen molecule

ozone molecule

Origin and action of ozone. The process begins when ultraviolet light strikes an oxygen molecule and splits it into its constituent atoms. The impact increases the momentum of the free oxygen atoms, one of which collides with another oxygen molecule and an atom of nitrogen. The result is an ozone molecule, which then absorbs ultraviolet radiation and is broken up into its components. This begins the process anew. The cycle absorbs the energy of ultraviolet rays and protects life on earth from their deadly radiation.

Large quantities of these artificial gases rise into the atmosphere every day from the millions of spray cans in use on earth. A vast and ever increasing number of household products come in such cans, and the latest findings of science appear to indicate that their gases do indeed have a destructive effect on the ozone. If so, the spray can might be regarded as the most dangerous end product of modern industrial society's criminal growth ideology.

In moving out to dry land, plants created the preconditions for animals, too, to leave the shelter of the seas and advance onto the land. The first pioneers were mostly arthropods—scorpions, spiders, and mites—which had evolved in the seas over enormous eons. Sea scorpions had reached lengths of around 6 feet. There were also sea centipedes, from which the land insects later evolved. But another 50 million to 60 million years were to pass before our remotest ancestors, the first vertebrates, of the Upper Devonian period on the verge of the Carboniferous, left the sea to live on land. That was about 350 million or 360 million years ago.

Who were these remotest ancestors, and what did they look like? Paleontologists can tell us quite precisely. In the first place, they know because vast numbers of fossilized vertebrates, primeval fish, have been found in the rocks of these geological periods. Then further confirmation came their way in 1938, when native fishermen in the western part of the Indian Ocean made a sensational catch of a fish that was remarkably primitive in appearance. It looked almost exactly like a fish previously known only as a fossil and thought to have been extinct for 350 million years. Because of its fleshy, fringed fins, it was called a crossopterygian (from the Greek *crossoi,* "fringe," and *pterygion,* "fin"). Here was a sort of living fossil, a specimen of an order of fishes from which the first land vertebrates had developed some 350 million years ago. The fishes of this group had a skeleton capable of evolving into the skeleton of a four-legged land animal. The fins gradually developed into short legs with five toes and the swim bladder into a lung, and the fish literally took the first step onto dry land. In this way did the first land vertebrate evolve out of the crossopterygian. It was a strange-looking amphibian, which scurried along on four legs, could breathe air, but in its external appearance strongly resembled a fish.

Well-preserved remains of this first land vertebrate—ancestor of us all—have been found in old rocks in Greenland. The animal bears the impressive scientific name of ichthyostega.

It is not clear why certain fishes—as well as certain plants and invertebrate animals before them—one day left their traditional biosphere, the unchanging environment of water, to crawl out onto land. Living in water was far less demanding. On land they had to contend with enormous variations in temperature that put a great strain on their metabolic functions; in addition, they exposed their bodies to the constant peril of drying out. Moreover, they had to fight gravity. Unlike organisms that live in water, land dwellers moving about on legs need some 30 to 40 per cent of their energy for the sole purpose of bearing their own weight.

Consequently, it was neither utilitarian nor probable for plants and animals to move out onto land. Abandoning their safe and reliable ancient environment must have involved a frightful risk for the more highly developed vertebrates, and it can be assumed that the move was not undertaken voluntarily. The venture onto land probably took place over long geological ages and was forced upon the organisms by a series of catastrophes. If our conjectures are correct, the situation must have strikingly resembled the oxygen dilemma that led to the development of breathing.

Our remote ancestors the crossopterygians, in their hunt for food, probably advanced into shallow coastal lagoons and rivers. The fossil record shows that during the Upper Devonian there were repeated spells of drought, which led to the periodic drying up of rivers and lagoons. The droughts undoubtedly caused the death of countless fish over millions of years by cutting off their retreat to the sea. Then one day fish with special physical characteristics developed the ability to survive out of water.

The ichthyostega and all the other early vertebrates that evolved from this fishlike amphibian remained partly dependent on their former environment for many millions of years. Their eggs, like those of modern amphibians, could develop only in water, and the hatched young had to pass through a tadpole stage, in which they breathed oxygen dissolved in the water through gills, as fish do, before they acquired the lungs and legs that permitted them to live on land. Each individual of the first land vertebrates, that is, had to pass through a brief summary of the long evolutionary process that had been necessary for fish to develop into land animals.

It is intriguing to think that we can all observe, in every pond populated by frogs and salamanders, the path followed by our ancestors some 350 million years ago.

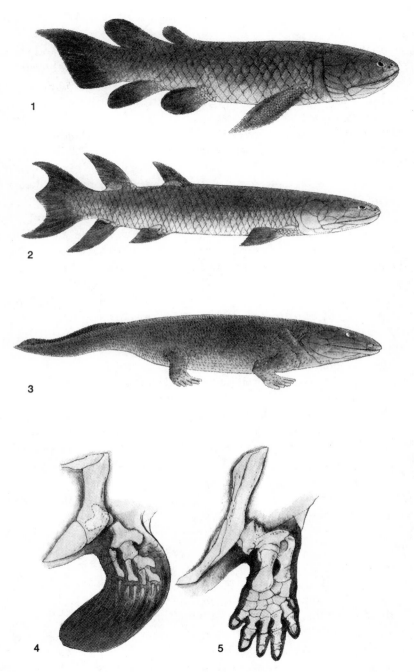

Evolution of the first land vertebrate, the four-footed ichthyostega (3), from the crossopterygians *Holoptychius* (1) and *Eusthenopteron* (2). (*Below*) Evolution of the skeleton of the forelimb in the first land vertebrates, *Eryops* (5), from the skeleton of a crossopterygian *Eusthenopteron* fin (4).

## The Climax of the Continental Biophase

In the Carboniferous period, so named from the vast beds of coal laid down then, the growth of vegetation on the continents probably reached its peak in quantity for the entire history of the earth.

The term scientists use for the totality of organic substances, for all the life forms existing on earth, is biomass. The weight of the biomass per unit of area can be estimated, and it serves as a kind of qualitative measure of the evolution of life.

In the Carboniferous period the vegetative biomass on the continents was probably higher than at any other period of geological history. Favored by a moist, warm climate, dense forests of giant ferns and horsetails spread over large parts of the globe, intermixed with scale trees, *Lepidodendron,* and *Sigillaria.* The lush development of land plants meant a further intensification of photosynthesis. The jungles of the Carboniferous period produced so much additional oxygen that the balance between consumption and production of oxygen, which had been temporarily maintained by the metabolic activity of marine plants, shifted more and more in favor of free atmospheric oxygen. Current theory holds that the oxygen content of the atmosphere 320 million to 380 million years ago far exceeded its present level.

But how can we know how high the oxygen level of the atmosphere was some 320 million years ago? Let us consider the question, at least briefly, because the answer affords a fascinating glimpse into the research methods and the collaboration of different scientific disciplines in solving problems of historical geology. Obviously, one can hardly expect to find unchanged samples of air—locked in the rocks, for instance—from such a remote epoch of the earth's past. The components of the air—especially the oxygen that is our concern here—are quite unstable and would promptly have combined chemically with the minerals in the rock. We owe our knowledge of the Carboniferous atmosphere chiefly to the fact that fossilized remains of gigantic dragonflies are embedded in the coal deposits that were laid down during this period. Imprints of dragonflies with a wing span of almost 30 inches have been found! These *Meganeura,* to give them their famous scientific name, were the largest insects the earth has ever seen.

In order to know how these giant insects could reveal anything about the oxygen content of the atmosphere, it is necessary to understand how an insect breathes.

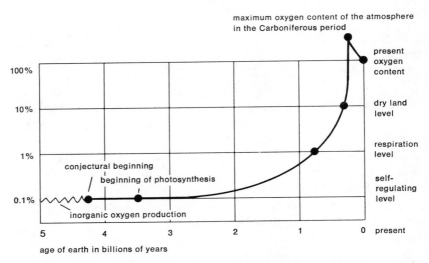

The increase in the oxygen content of the atmosphere over the course of geological history from the origin of the earth to the present (adapted from M. Schidlowski). The age of the earth is given in billions of years, the atmospheric oxygen content in percentages, with the present oxygen content taken as 100 per cent.

Insect respiration differs from that of other animals, such as birds and mammals, in that insects do not actively inhale air into lungs by means of the appropriate muscles. Instead, the vital gas penetrates into their bodies of its own accord. It does this by way of tiny, widely ramified tubes—the tracheae—which carry oxygen to every organ and cell. Simple atmospheric pressure drives the oxygen molecules through openings in the insects' chitin armor and into the network of fine tubes. In most species of insects, the penetration and distribution of the oxygen molecules are aided by normal bodily movements or by a rhythmic contraction of the abdomen—which alternately compresses and expands the tracheae.

This passive type of breathing has one rather important consequence: it is so inefficient that it sets limits to the size of insects. Today's atmosphere does not exert enough pressure to supply tracheal tubes any longer than those insects now possess.* That was not so in the Carboniferous period. The giant dragonflies, with their long

---

* For this reason, the giant insects of science-fiction movies, which crawl over the earth devouring people and railroad cars, would promptly suffocate in the earth's atmosphere. The spiders frequently shown in such films are not insects, incidentally, but they would suffer a similar fate. If the otherwise eminently

tracheal tubes, could have survived only in an atmosphere with a much higher concentration of oxygen than exists at present. That is how, through our knowledge of respiratory techniques, we have been able to deduce the oxygen content of the atmosphere.

The jungles that spread over much of the world during the Carboniferous period flourished in the midst of extensive swamps. The swamps had formed in huge shallow hollows and basins that resulted from the upwarping of mountains. The mountains were not nearly as high as the Alps, the Rockies, or the Himalayas of the present day, and so rain clouds from the oceans could reach almost any region of the continents unhindered and drop their freight of water. The abundant rainfall led to lush growth of vegetation, accompanied by enormous oxygen production from the increased plant photosynthesis.

The released free oxygen could, of course, enrich the atmosphere only if the organic, carboniferous plant material was not completely reoxidized.

An example will make this clear. An apple tree, say, makes carbohydrates with the aid of photosynthesis. The carbohydrates are chiefly used for the growth of its foliage, branches, and trunk. Also, over and above the needs of its own metabolism, the tree converts some of the carbohydrates into fruiting bodies, or apples. We eat the apples and incorporate their carbohydrates into our own bodies. With the aid of the oxygen we inhale, we can then convert the carbohydrates back into the substances from which they were originally made, water and carbon dioxide. Every day we similarly assimilate the carbohydrates of many types of plants, not only in the form of fruit but also as leaves, such as those of lettuce, cabbage, spinach, and so on.

When we cut down the apple tree and burn its wood in a fireplace, the flame makes visible this same process of oxidation, the same release of chemical binding energy, that takes place in a slow and controlled fashion in our bodies. Even if the tree were slowly to die a natural death, its entire organic, carboniferous substance would gradually be oxidized, would combine with the very oxygen that the tree had once released. In the end, all the oxygen released by photosynthesis would have been used up and would once more be bound. The oxygen cycle would be complete.

---

successful insect class had found a better solution to the respiratory problem, by developing an active breathing musculature and lungs, it would undoubtedly dominate the earth today, rather than we human members of the phylum of vertebrates.

The situation was far different in the Carboniferous period. For one thing, there were relatively few land organisms in the enormous primeval jungles to eat the plants and consume their released oxygen. As plants died, almost all the organic, carboniferous substance of the trees and undergrowth sank into the bottomless morass of the swamps, where atmospheric oxygen could not get at them. They were withdrawn from the process of reoxidation. By complicated biological and chemical processes, the vegetable substance of these plants was converted over the course of ages into coal. Free oxygen increased only because so many organic carbon compounds were buried in the sediments of the earth's crust.

During the Carboniferous period another important event occurred that we must mention briefly. The first reptiles evolved out of the first land vertebrates, the amphibians. The amphibians could reproduce only in water, so that every generation had to conquer the land anew, so to speak. But the reptiles represented a class of vertebrates that in

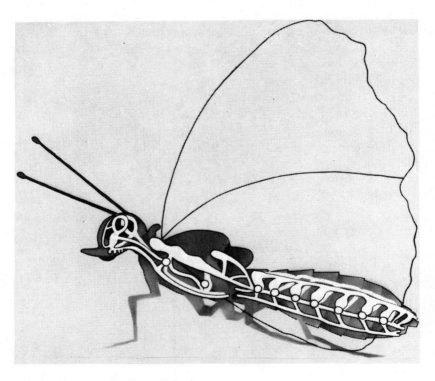

Tracheal system of an insect's body

this respect had become independent of water, the environment from which their ancestors came. They owed their independence to the type of egg they had developed. Reptilian eggs could be laid on dry land; they were protected from drying out by their thick, leathery skin. The inside of a reptile's egg, however, is a reminder of the watery environment that the eggs of the more primitive amphibians required for their development. The reptile's egg carries its own small pond within it. And before the young reptile hatches it passes through a kind of tadpole stage in this miniature pond. But by the time it hatches it has exactly the same shape as its parents and is equipped with all the organs it needs for living on land. The reptiles of today, such as lizards and snakes, still lay the same kind of egg that was developed by their ancestors some 300 million years ago. What is more, similar eggs grace our breakfast tables every morning and lie in the nests of birds. The birds, which evolved from reptiles, took over their reproductive system and developed it further.

In the geologic periods that succeeded the Carboniferous, reptiles developed into more differentiated and ever larger forms. For more than 200 million years they were the dominant class of animals on earth. Like the birds and mammals of today, hundreds of different reptilian species adapted to all imaginable biospheres. In the Mesozoic era, from 230 million to 70 million years ago (the era is subdivided into three periods, the Triassic, the Jurassic, and the Cretaceous), the evolution of reptiles reached its climax in the form of the giant saurians—the largest land creatures of all time. The biggest of them, the herbivorous brachiosaurus, walked on four legs and reached a length of 75 feet and a weight—inconceivable for a land animal—of 80 tons. These colossi lived half underwater—much like the modern hippopotamus—in rivers, lakes, and swamps, so that the water could help to support the weight of their enormous bodies. Because of their size and their long necks, the brachiosaurs could graze while standing in water more than 30 feet deep. They probably left the shelter of the water only to lay their enormous eggs.

Such a body size posed special problems for the giant saurians. Excavated skeletons of the largest species indicate that these huge creatures probably had a kind of second brain near the base of the tail. It served as a sort of relay station for reinforcing the signals sent out by the distant brain in the animal's head.

While it is probable that the climax of the vegetative biomass on the continents was reached in the Carboniferous period, the multiplication of the biomass as a whole, both plant and animal, most likely peaked during the period in which the great herds of saurians dominated the continents. In quantitative terms, that is, the evolution of life reached its climax not in the present but, in all probability, 100 million to 200 million years ago, in the Mesozoic era.

It should be noted that very early in the Mesozoic era the direct ancestor of us all, the first warm-blooded mammal, which was not much bigger than a mouse, evolved from small reptiles.

## The Desert Catastrophe

Among the lofty deserts of the Colorado Plateau there is a unique primeval forest—although the word *forest* does not entirely describe it. When I first saw the place, I had the impression that giants had plucked up trees by the handful, like weeds, and dropped them on the ground. Their trunks, often piled haphazardly one on the other, covered the desert floor as far as the eye could see. The thick roots and branches of the trees were all broken off, and even most of the trunks were smashed into chunks of different lengths. The ground among the trunks was littered with scraps of wood of every size, and with bark in which the tunnels of wood-eating insects could be seen. In some of the tunnels there were still beetles and larvae.

Although not a soul was to be seen far and wide, the forest looked as if it had just been felled. You imagined that you could still hear the growl of chain saws and the blows of axes. In fact, total silence prevailed, except for the occasional slight rustling of the desert wind among those mighty stumps. The whole scene had something ghostly and weird about it. The mood in that felled forest was like the atmosphere on the day after a natural catastrophe.

But the most extraordinary thing about these shattered forest giants was that all of them without exception, including the scraps and the bark, were made not of wood but of colored precious stones, agate, chalcedony, opal. The gems gleamed in the loveliest of colors, especially along their planes of fracture.

The trees are 200 million years old. They date from the Triassic period of the Mesozoic era. Before they were uprooted by some storm or flood, saurians grazed among their trunks. As the fallen trees lay buried for 200 million years, ground water trickled through them.

13. In the sequoia forests of the rainy Sierra Nevada in California

14 and 15. Diagonal and cross-stratified ledges of sandstone on the Colorado Plateau, the petrified remains of enormous desert dunes of the Mesozoic era

16. Petrified footprints of a giant saurian that walked here more than 100 million years ago. On the northwestern rim of the Sahara.

17. Its tough dead leaves shade the madar tree during dry periods. On the horizon, a very localized rain is falling after years of drought.

The minerals in the water filled the cells of the wood, petrifying the trees and preserving them down to the present. Because petrified wood loses its elasticity, the trees were splintered by the movement of the rocks in which they were embedded. Many saurian skeletons also have been preserved in nearby rock formations.

It is not exceptional for the remains of trees and saurians to have been found in the strata of the Colorado Plateau, in what is now largely a desert landscape. Rather, it is highly characteristic of geological history that fossilized saurian skeletons have most often been found in the midst of deserts and parched steppes, or at any rate in drought regions. Perhaps the most famous sites of all are in the arid steppe of East Africa, the Gobi Desert, the deserts of Utah, and the Sahara.

I became most keenly aware of the climatic changes that have taken place since that peak of organic evolution several hundred million years ago the first time I literally walked in the gigantic footsteps of a saurian. In the midst of one of the driest wastelands of rock on earth, I stepped into the footprints of an animal that could have lived only in a humid, tropical climate where vegetation grew lushly.

Footprints of the saurians, like their skeletons, have in some places been preserved intact. These are prints that were pressed by the animals' enormous weight deep into the soft, muddy bottom of drying bodies of water. Some of the prints are so sharp that we can clearly discern the bulges of mud that squished up between the animals' toes. Usually the tracks were preserved because the empty riverbeds or basins of lakes did not immediately fill again with water, but lay drying for such a long time that the mud—laced with deep, narrow cracks—was gradually baked by the sun to the hardness of stone. Otherwise, the tracks and cracks would have softened and disappeared.

Long periods of drought between rainfalls withered the vegetation near the drying lakes and rivers, so that more and more bare ground was exposed. When a heavy rainfall came at last, large quantities of disintegrated stone and sand were eroded and washed into the lake

18. Dramatic lighting in the Arizona desert as a snowstorm approaches

19. A young peregrine falcon surveys the scene.

20. Female of the spotted sand grouse (Pterocles senagallus) turning the eggs of its clutch. A cover stone can be seen beside the bird.

basins or riverbeds. Before the baked mud could soften, the concavities of the tracks and cracks were filled with a thick layer of sand and rock dust and covered over. As time passed, more and more layers covered the tracks and ultimately solidified into stone. During the Cenozoic era (from the Greek *kainos,* "recent," and *zoon,* "life") these deposits were worn away again by weathering, so that the ancient stratum of mud in which the saurian tracks had been imprinted emerged once more into the light of day.

Among the most fascinating saurian tracks preserved in this fashion are the gigantic footprints of the mighty 80-ton brachiosaurus of North Africa. The prints are approximately 3 feet in diameter (see Plate 16). The fact that jungles once flourished and saurian herds grazed in what today are almost lifeless deserts indicates that those deserts must have spread steadily toward the end of the Mesozoic era and the beginning of the Cenozoic.

In other words, no sooner had plants and animals adapted to life out of the seas than the desert began once more to assert its sway over the continents. We know that this development began in the Mesozoic era—the geologic Middle Ages—because a desert from that era is still extant, the fossil desert on the Colorado Plateau, which arose in the Jurassic period, between 160 million and 180 million years ago. It is one of the most impressive landscapes in the western half of North America.

The fact that the massive strata of rock—as much as 2,000 feet thick—in this fossil landscape consist of tilted and cross-stratified sandstone beds permits the conclusion, supported by other geological data, that the stone was once enormous sand dunes heaped up by the wind (see Plates 14 and 15).

Dipping strata are formed when material—for example, sand—has been transported chiefly in a horizontal direction. A current of wind blowing steadily in one direction forms ripples and dunes on sandy soil. Along the flat windward slope of the ripples and dunes, sand is constantly carried away by the wind, and it piles up again on the steep lee side. Stratification occurs when first the large, heavy grains of sand are deposited in the lee and then the small, lighter particles. If there are several successive depositions, sharp strata boundaries are formed where the new layer of coarser material is deposited on the finer material from the preceding deposition. Cross-stratification is produced when a change in wind direction creates an inclined ripple or dune running at an angle to the dune or ripple beneath it.

Since inclined and crossed stratifications can also be produced by currents of water in the sandy deposits at the bottom of an ocean or river, other indications in rock are needed before the geologist can definitely identify petrified dunes. The inclinations produced by wind ripples have a different angle of dip from those produced by water ripples. Also, grains of sand transported by wind are much more rounded than grains of sand transported by water.

Skeletons of saurians are embedded in the vast petrified desert dunes of the Colorado Plateau. It may be assumed that the saurians of this region either starved to death or died of thirst as their environment turned increasingly into desert.

The renewed spread of deserts in the Mesozoic era can be demonstrated on a world-wide scale. The Sahara probably originated then. Fundamentally, the desert had never abandoned its claim to hegemony over the continents. It had merely been confined and thrust back into smaller areas during the tempestuous time when life was evolving on dry land. Even in the Carboniferous period, with its vast jungles, there were small desert regions, as rock strata of the period indicate.

When these deserts gradually began to expand again during the Mesozoic era, it was like the outbreak of an old continental disease that had never quite been cured. And the "relapse" had one very serious consequence: the spread of the deserts threatened the balance that had been achieved between oxygen production by plants and oxygen absorption by the rock of the crust. The land plants had

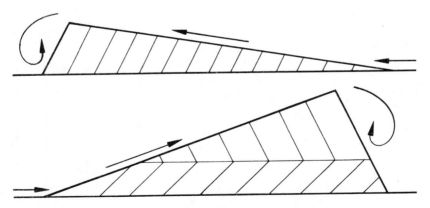

How inclined strata and cross-strata are created

shifted the balance toward an atmosphere richer in oxygen. Now the shift was the other way, and it appeared likely that the oxygen content of the air would slowly diminish.

In this remote period of 160 million to 180 million years ago, the deserts did not begin a regular advance, spreading farther and farther across the continents right up to the present. On the contrary, our planet underwent extreme fluctuations in climate. Large parts of the continents, including the present Sahara region, sank beneath the waves and were covered by the sea for many millions of years. World-wide phases of tropical climate occurred, but ice ages also intervened, amid the predominantly warm climate of the planet.

On the whole, however, desert recurred more and more frequently, and over ever larger areas; in the recent geological era, we find that more than a third of all the continental land area has reverted to desert or desertlike landscape.

What are the causes of the slow, frequently suspended, but apparently inexorable march of the desert over the continents that had once been so successfully conquered by plants and animals?

Before I turn to the fascinating data scientists have amassed about this vital question, I must describe a mysterious, almost incredible experience I had in 1965 in the North American desert.

# 3 / Parching Continents

## The Secret of the Solitary Valley

The track we were driving on resembled a dry stream bed filled with stones and gravel. Certainly it bore little resemblance to a passable road. There were stretches where we could go no faster than a walk. Even worse were the steep climbs, which often went on for miles. Then the wheels of our vehicles would churn fruitlessly in the loose gravel, and we made no progress at all. The spinning tires, heated to the bursting point, hurled stones as big as a fist in all directions. We got stuck several times, and only by exhausting effort were we able to get our two vehicles, overloaded with water and gasoline, provisions and equipment, moving again. But eventually we learned how to steer the vehicles so that the wheels rode over stones that were neither too small nor too large.

Our route was occasionally blocked by sizable boulders, and occasionally by wild burros. These were descended from the donkeys of gold prospectors who had passed through this part of the Mojave Desert. The boulders we had to move aside by brute force, but the shy burros quickly scattered of their own accord.

We were surrounded by the grandiose, chaotic panorama of a mountain desert. The rocks of the mountain range had been worn away, over millions of years, by extreme alternations of temperature and by the sandblasting of the desert wind. The mountains, from 6,500 feet to nearly 10,000 feet high, seemed to be drowning in the rubble produced by their own weathering. The scree looked scorched by the sun's rays. Obeying the law of gravity, it had slid slowly into the valley, a gigantic glacier of stone. The somber landscape was relieved by the brilliant yellow blossoms of the fishhook cactus (so called because of its hooked spines) growing in the midst of the black and red sun-warmed detritus of rock.

Our destination was a remote mountain valley, and we still had some 25 miles to go. A mysterious phenomenon, unique on earth, was said to occur in this valley. Before starting on our expedition, we had tried to find out more about it, but the ranger whom I queried—I knew that on his patrols he sometimes passed through this remote area—had skillfully contrived to evade my leading questions. Finally I had to ask him bluntly; only then did he reply, "Yes, yes, they're still there." I could see that he gave the information with extreme reluctance, and I had the impression that the minute he spoke he was sorry he had. He said that at "Teakettle Junction"—a fork in the track named after an old marker, a teakettle stuck atop a wooden post—we should keep to the right, and that under no circumstances must we spend the night in the valley.

Years later, during a second stay there, I found out one reason why the rangers tried to keep strangers from visiting the valley, although they could not directly forbid it. The region was so out of the way that it made an ideal trading post for drug smugglers. The authorities had discovered that drugs by the ton were brought in from Central America in recklessly daring flights of four-engine planes. The pilots evaded radar controls by flying low between the wildly crevassed mountains of the Mojave Desert. If we had known about the smuggling at the time of our first expedition, we would certainly have followed the ranger's advice and not spent the night in the valley.

Rarely have I been so filled with suspense on a desert journey. After we passed Teakettle Junction, late in the afternoon, the ridge of hills allowed us our first glimpse of the mysterious valley. We stopped at once and began to scan the valley with our binoculars, but without finding what we hoped to discover. We were still too far away to make out details. The valley did present an unusual appearance. It formed a large circular flat, almost white in color and several miles across. An ideal natural airfield. This natural amphitheater was surrounded by high black mountains, from the foot of which enormous fans of rubble extended for miles to the edge of the white flat. The valley bottom consisted of mingled salt and clay deposits, composed of the finest particles, from the weathering of the surrounding mountains. Rare but torrential rainfalls washed the salt and clay particles out of the fans of waste and floated them to the center of the valley. There the water evaporated rapidly, while the finely weathered material built up, layer on layer, over millions of years. This type of sedimentation builds up what is known as a playa. The surface of the salt and clay that are left after the water evaporates shows not the slightest slope—the water acts as a natural level. That feature is of the greatest significance in the story I am about to tell.

Darkness had set in by the time we reached the edge of the playa. During the night I was awakened by a curious noise, a distant rumble interrupted by hard, rattling noises. It must have been due to an avalanche of stone plunging thunderously down the steep flanks of the mountains into the valley.

Next morning, just as the shimmering white disk of the sun was rising above the ridges in the east to bathe the white flat in glittering light, we set out on foot to explore our environs. The dazzling white ground we walked on was hard as stone, but was covered with a dense network of deep cracks caused by contraction while drying. We had been tramping about half an hour when I saw, directly in front of me, a long dark line on the surface of the playa. That was what we had hoped to find in this solitary desert region. The line appeared dark only because of the counterlight; it was actually a slight depression in the ground, a curious track that stretched on for miles and looked as though it had been left by some heavy object dragged across the hard surface. In one direction the track went on until it merged with the glittering brightness of the playa. In the other direction the track ended in a small dark point, though at a distance hard to estimate. That object, whatever it might be, seemed to have left this track behind it. Full of eager expectation, we hastened along the track toward that distant dark point.

It was strange; although we were walking as rapidly as we could, at times jogging, the mysterious object came no nearer. The distance between it and us seemed to remain always the same. We had the feeling that it was moving away from us at the same speed at which we were pursuing it. Undoubtedly an optical illusion was involved, for in this perfectly flat desert, without a trace of vegetation, we had nothing to orient ourselves by, nothing by which to estimate distance.

When we had covered about a mile and a quarter, the distance between us and the dark object suddenly began to diminish rapidly. The long track ended in a large, angular rock. There could be no doubt that this was the object that had made the track. But the block of stone obviously weighed between 400 and 600 pounds. What force had propelled it? No matter how hard the two of us tried, we did not manage to budge the boulder by so much as an inch.

Later, when we found smaller rocks that we could turn over or pick up, we saw plainly that their undersides were highly polished from years, or perhaps thousands of years, of migrating across the floor of the desert.

Although many investigations have been made over the years, no entirely satisfactory explanation has ever been found for the movement of those boulders and stones across the perfectly level surface of the playa. Indeed, the motion of the "moving rocks," as they are called, seems to vary a great deal. Measurements of the position of stones and the length of their tracks over a span of many years have shown that the big boulders move only about three feet or so a year, and that for years at a time they do not advance at all. Smaller stones, many no bigger than one's fist, seem to migrate a good deal faster. As far as I know, no human being has had the good fortune actually to see the mysterious motion of one of these rocks or stones.

## The Onion Model of Desert Formation

If scientists know little about what makes tiny portions of the earth's crust—the moving rocks just described—travel across the desert floor, geologists and geophysicists have learned a great deal about the natural laws to which the entire crust is subject, and which have been leading to the increasing devastation of the continents.

As we saw in the last chapter, the crust, which we tend to regard as the fixed, immovable substructure of all our activities, is in fact in constant movement. Recurrent reports of disastrous earthquakes constantly remind us that we live on the surface of a thin skin of rock covering a vast and restless molten sphere. The seemingly rigid rock of the earth's crust can be forced out of shape by pressures originating in the interior of the earth; and such deformations contribute to the creation or extension of deserts. A unique example, again in the western part of the United States, provides evidence of that.

In the southeastern corner of Utah, on the part of the Colorado Plateau that extends into that state, is a bulging rock formation some 2½ or 3 miles across. In the middle of this enormous boil of rock yawns a deep crater—an abyss 3,000 feet deep whose interior is an

21. The author engaged in geological studies in the Sahara. The spherical stones are lobolites, the fossils of marine organisms.

22. One of the moving rocks of the Mojave Desert

extreme desert. The rock bulge was created by the pressure of a tremendous salt dome. A period of desertlike climate almost 300 million years ago caused the evaporation of a great inland sea in North America. It left behind a layer of salt 3,000 feet thick. In the further course of the earth's history, a wide variety of sedimentary rocks, which today form the Colorado Plateau, were deposited on top of the layer of salt. Under the tremendous pressure of mineral strata as much as 6,500 feet thick, the salt deposit assumed a plastic state. Wherever the rock had been thinned by weathering, the salt, rebounding from this inconceivable pressure, heaved up the strata of rock and penetrated underneath them to form a salt dome. The area surrounding the salt dome slowly sank as salt flowed into the bulge.

The manner in which the rock strata were deformed by pressures from below, and the way the crater desert was created in the heart of the bulge, can best be demonstrated through the example of an ordinary onion. With a large round onion and a knife we can re-enact this principle of desert formation.

First the onion is cut in half, as indicated in the illustration below. The upper half of the onion represents the bulge in the rock formations created by the upward pressure of the salt dome.

Now trim away the upper third of this onion half, as shown in the next photograph. The cut-off piece represents a massive stratum of rock that in the course of time has been worn away by weathering.

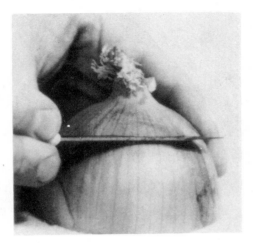

Next, to see how the rock strata in the interior of the bulge are deformed, how the various strata are arranged, and how the crater desert came into being, cut through the remaining two-thirds of the onion half, this time vertically.

The shell-like layers of the onion show by their arrangement how the horizontally stratified layers of rock were bulged out and deformed by the pressure of the salt.

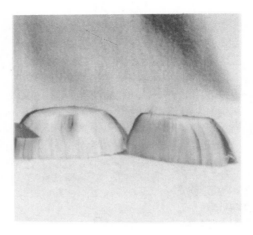

Since the inner "shells" of rock consist of stone that is softer than the stone of the outer shells, the wind and water that penetrate through cracks and crevices in the outer layers wear away the inner layers. A hollow results. That process can also be demonstrated, by removing the inner shells from both our remaining pieces of onion, as in the photograph.

When we place the two pieces together again, we have a completed model of the rock bulge and the crater desert. The shells of the onion can be clearly seen as rings. In the same way, the preserved shells of rock form several rings around the deep crater desert.

Shielded from cooling air currents by these high exterior shells of rock, the air masses in the interior of the crater are heated by the intense solar radiation of the summer months to temperatures as high as 50° C. (122° F.). The rocks at the bottom of the crater can even reach 80° C. (186° F.). If an occasional rain cloud from the distant Pacific ever manages to reach the parched Colorado Plateau and climb the flanks of the rocky bulge, the rain strikes the cauldron between the crater walls and evaporates at once.

This remarkable example of how deformable the earth's crust can be also serves to illustrate how a relief desert is created. The crater desert owes its origin to specific and rather rare conditions. But the area in which it is situated—indeed, the whole western part of the North American continent—has largely turned into desert because the earth's crust has so distinctive a relief there. The altitude is either very high, like that of the Colorado Plateau, or very low, like that of the narrow valleys of the Mojave Desert. Both types of landscape are surrounded by high mountains, which prevent clouds from the Pacific from reaching them. The same conditions prevail in vast areas of Central Asia.

Both continents are affected by an additional factor: their interiors are a great distance from the sea, and rain clouds cannot travel that far. The clouds dry out in the course of their long journey. That unfortunate climatic effect is characteristic of all large continents. It is particularly noticeable in wide regions in the center of North America east of the Rocky Mountains. It is also applicable to such a subtropical desert as the Sahara, which is continental in its proportions. If clouds ever do enter such a region, they rarely survive all the long way to the center.

While subtropical deserts can form only within certain latitudes (see chapter 1), almost any region on earth can become a relief desert. If we wish to understand the earth's past and its future fate as a desert planet, we must pause here to ask how the high plateaus, the deep basins, and, above all, the mountain barriers at the edge of the continents first came into being. The answers to many of these questions are still being sought by scientists working in highly specialized fields, and our review of their findings will of necessity be considerably simplified.

## Floating Continents

The solidified crust of the earth, which is so thin in proportion to the earth's total mass that we can properly speak of it as a skin of rock, is not wrapped around our planet as flawlessly as the skin on a peach. It consists of several individual slabs, or floes, which geologists call plates. These are traversed by innumerable cracks. By far the largest part of the earth's skin of rock—around 70 per cent of it—is covered by ocean. The remaining 30 per cent forms the visible part of the crust—the continents and such splinters of them as Greenland, Madagascar, and New Zealand.

The vast plates cannot be delineated as either purely oceanic or purely continental portions of the crust; they include more or less extensive areas of both types of crust. The North and South American plate, for example, is composed of the oceanic crust of the western half of the Atlantic and the continental crust of the American mainland.

The floor of the oceans consists chiefly of heavy basalt. For the most part, the continents are made up of relatively lighter rocks, such as granite. Like icebergs that float deep in the ocean, with only their tips visible, the "light" continental plates float on the viscous, molten magma of the earth's interior. They sink into the magma to depths of about 18 to 45 miles.

As we saw in the case of the coral reefs in the desert sands, the present distribution of land and sea is only a phase in the long history of the earth. Large parts of the continents may someday be temporarily flooded, because of crustal sinking or a rise in the sea level; and then they may go through the reverse process and dry out again,

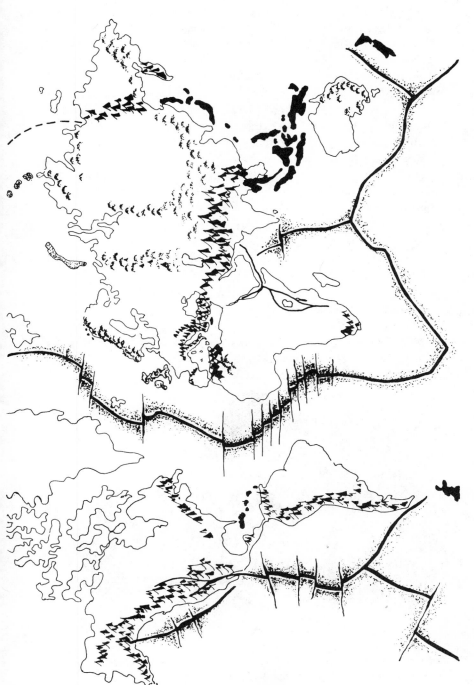

World map of the crustal plates, showing also the mid-oceanic trenches and ridges and the continental trenches and mountains

because of a lowering of the sea level or an elevation of the crust. Regardless of random changes in coastlines, however, the distinct division of the crust into high continental sections and low oceanic sections is a constant phenomenon of our planet. No matter how far the sea may sometimes flood into the margins of the continents, the average continental altitude, 2,800 feet above sea level, stands in sharp contrast to the average depth of the crust beneath the sea, 12,400 feet below sea level—a difference of more than 15,000 feet.

The comparison of the continents with floating icebergs is made even more apt by the fact that the continents floating in the magma are not fixed to one spot on the globe, but drift and constantly change their positions in relation to one another. In looking at a map of the Atlantic, showing Europe and Africa to the east and North and South America to the west, we cannot help noticing how well the western coast of Africa and the eastern coast of South America fit together. The impression is reinforced if we use a map that shows not only the present coastlines but also the real limits of both continents—the edges of their continental shelves. The continental jigsaw puzzle can be enlarged to include the other southern continents, the Antarctic and Australia, and the Indian subcontinent. If these continents are pushed together, they form, along with Africa and South America, one almost seamless giant continent.

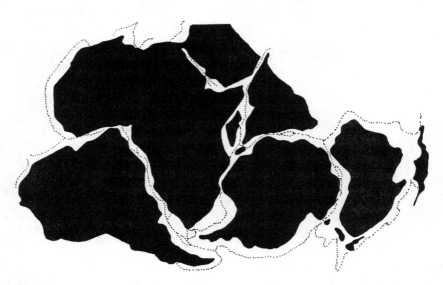

The single giant continent of the Southern Hemisphere (Gondwanaland) as it must have been before it broke apart at some time in the earth's past

The continents and the one subcontinent of the Northern Hemisphere—North America, Greenland, Eurasia—can likewise be fitted together. Of course, it is possible that such a fit is a matter of pure chance. However, working from the premise that all the southern continents may once have formed a unit, geologists have discovered that rocks on both sides of the Atlantic, in Africa and South America, coincide, and frequently are identical down to the smallest detail. So close is the agreement that very specific structures are continued on the opposite continental shore. Another discovery has proved even more dramatic and important: on all the southern continents, including the subcontinent of India, fossils of the same species of animals and plants have been found. These species must have occupied a coherent land area, for otherwise they would have developed into independent species, as in fact happened later, after the giant continent split asunder.

Scientists have by now gathered so many details about the structure of the earth's crust that the concept of drifting continents is generally accepted. The old southern continent and the old northern continent probably split into parts in the Mesozoic era, about 130 million years ago—in the age of the saurians. A new ocean, the Atlantic, formed between Africa and South America and between Europe and North America. We are witnessing a similar process today. In our present geological era, the Cenozoic, the huge African crustal plate is breaking up, along deep fault lines, into two parts, which are drifting away from each other. A number of large lakes—Lake Rudolf, Lake Tanganyika, and Lake Nyasa among them—have formed along the fault, making it easily traceable on maps (see page 103).

This East African trench system, as it is called, extends northward along the Red Sea, the Dead Sea, and the Jordan Valley all the way into Lebanon, a distance of several thousand miles. In the Red Sea it can be seen that the fracture and drifting apart of the two segments of the continental crust, which is taking place at a rate of about 1.6 centimeters per year, has resulted in the almost complete separation of Africa from Arabia. The penetration of sea water into the crustal fault, which averages 155 miles wide and 6,500 feet deep, has already created an embryonic ocean, the Red Sea. Here a continental fault line—which it still is in East Africa—has resulted in an oceanic fissure system. Comparable fissure systems run through the bottom of all the oceans. Along with the continental fault lines, they form a world-wide system of crustal dilation (see the map on page 103).

The ascent of juvenile magmatic material into the oceanic fissure systems results in the steady formation of new oceanic crust, which is observable as mid-oceanic ridges on both sides of the great oceanic trenches. A consequence of this process is that the ocean floor keeps expanding. Fresh ocean floor is constantly being attached to the continental crusts as they drift apart. The internationally accepted term for the process is sea-floor spreading.

The continental fault lines, which within relatively short geological time spans are transformed into oceanic fissure systems, represent the initial aspect of the trenches responsible for continental drift, crustal growth, and ocean formation.

The crucial motive force for the drift of the plates is probably not

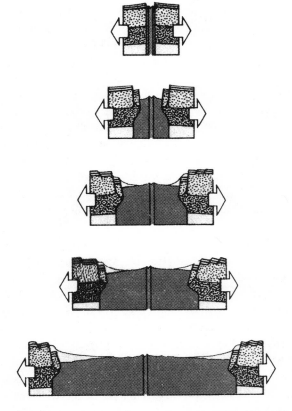

Several phases in the spread of oceans: (*top to bottom*) formation of a continental trench (East Africa), the Red Sea, the Gulf of Aden, the early stages of the Atlantic (after Heezen, 1969)

the magma rising into the fissures. Rather, the force probably arises from circular upward convection currents originating in the viscous interior of the mantle. These currents tug at certain weak spots in the continental crustal sections, such as may be observed today along the East African fault, pulling the plates apart and starting them moving. The resultant fissures are constantly being filled and healed by the molten magma that rises from the earth's mantle, steadily producing new oceanic crust, and the new crust, by its sideward pressure, contributes to pushing the plates apart.

The convection currents are the result of differences of temperature in the earth's interior. Such currents are essentially movements tending to balance the distribution of heat within a volume of matter. The rising and falling of the tea leaves in a cup of hot tea are a visible demonstration of the pattern of convection currents.

Earthquake studies have shown that the mantle lying beneath the solidified crust also is predominantly solid. The rock of the mantle, however, is under extremely high pressure. The high temperature keeps it in a viscously plastic condition and permits a heaving motion of a few centimeters a year.

The Atlantic Ocean represents an advanced stage of continental drift and the formation of oceanic crust. The bottom of the Atlantic is split by a deep trench running north and south along its entire length. On both sides the trench is flanked by the massive new crustal formations of the Mid-Atlantic Ridge. This new oceanic crust, constantly fed by undersea volcanoes, rises above the surface of the water in a few places, in the form of volcanically active islands. The best known of these is Iceland, the island of glaciers and volcanoes. Farther to the south, the Azores, Ascension Island, St. Helena, Tristan da Cunha, and Bouvet mark the position of the Mid-Atlantic Ridge.

It is also striking that the line of the central trench largely coincides with the edge of the continental shelves of Europe and Africa on one side and those of Greenland, North America, and South America on the other side. If, in an imaginary jigsaw puzzle, we push the continental plates toward each other along the Mid-Atlantic Ridge, we will have the starting position from which the land masses drifted apart.

Since the continual formation of new crust has not caused the earth to swell up like a balloon filling with air, the world-wide oceanic zones of active crustal formation must be balanced by a comparable world-wide system of structures in which crustal reduction is taking place. And, since the zones of oceanic crustal formation are located where rising convection currents bring material up from below, the corresponding zones of crustal reduction must be in areas where the currents are moving downward. Such zones in fact exist, and in places are approximately 6,000 miles apart.

What this means is that every major plate has on one side a line along which the crust is increasing and on the other side a line along which the crust is decreasing. It is this balance that determines the overall direction of the plate's movement.

From its hot birthplace along one of the mid-oceanic trenches, the plate moves toward a zone of crustal reduction, cooling as it moves and so becoming heavier, until finally it surpasses in density the matter of the mantle and sinks down in the mantle along the direction of a downward convection current. After such a migration, lasting millions of years, the solidified crustal material eventually reaches a region so hot that it becomes molten again.

The crustal movement from zones of build-up to zones of breakdown is like that of material on an endless moving belt. The lighter, higher continental portions of the plates are carried along in the movement. The zones of rising convection currents and crustal build-up are marked either by a fault line on a continent or by a mid-oceanic ridge. The zones of downward convection currents, or crustal breakdown, are frequently marked by what geologists call a geosynclinal trough, in the form of a V-shaped deep-sea gully adjacent to a high continental folded mountain range. The genesis of these deep-sea gullies is especially clear in the case of the one on the west coast of the North and South American plate, which is drifting westward.

There is a mid-oceanic ridge in the Pacific, west of the South American continent, that parallels almost the entire length of the continent. Along this ridge active crustal formation is taking place. Part of the ridge is missing west of the North American continent; apparently it has been overridden by the continent in its westward drift.

One of the two plates in the Pacific Ocean is drifting eastward, in the opposite direction from the American continental plates. On the western rim of the Americas, this eastward-moving plate follows the downward direction of convection currents there and dips beneath the continental plates. The result is a geosynclinal trough just off the coast of the Americas.

The high mountains that run along the entire Pacific coast of the Americas, from Alaska to Tierra del Fuego, resemble a gigantic bow wave being pushed along by the westward-drifting continents of

North and South America. The mountains mark the western limit of the great continental plates. They also mark precisely the place where two convection currents in the mantle descend into the depths of the earth's interior, carrying oceanic crust with them.

One might well wonder how the process of swallowing the earth's crust could lead to the rise of the continental crust and the formation of mountains. In fact, these two opposed processes are complementary, not contradictory.

It is important to note that the North and South American plates have apparently drifted so far to the west that their mountainous western rim lies directly over the zone of downward-moving convection currents. This would seem to indicate that crustal reduction is somehow connected with mountain formation.

Remember that the continents are relatively light in weight. The density of their rocks is so low, compared to the density of the magma in which they are floating, that they cannot be carried downward by the convection currents and swallowed up by the earth's mantle. Like pieces of driftwood floating in a stream, pieces of continental plate collect where the movement of the heavy oceanic crust is directed downward—but the continental materials stay on top. For this reason, many scientists assume that pieces of continental crust pile up and cleave together to form mountains. In other words, the crust thickens. But such a process does not of itself fully explain the existence of high folded mountains at the edge of continental plates.

The processes that have led to mountain formation in the past and will do the same in the future are among the most complicated geophysical events on our planet. Dozens—in fact, hundreds—of different processes contribute to the formation of mountains. The least complicated type of formation results from the collision of two continental plates. It may be compared to the consequences of a collision between two automobiles.

We are witnessing such mountain building today as the result of crustal collision. A fragment of the old southern continent that is drifting northward—the Indian subcontinent—is colliding with the gigantic plate of the Asiatic continent. The "zone of crumpled fenders" of this continental collision consists of crustal plates that have been forced up to heights of nearly 30,000 feet—the enormous rocky folds of the Himalayas.

On the western edge of the two American plates, especially of the South American one, the dipping of the eastward-drifting Pacific plate under the westward-drifting continental plates is not proceed-

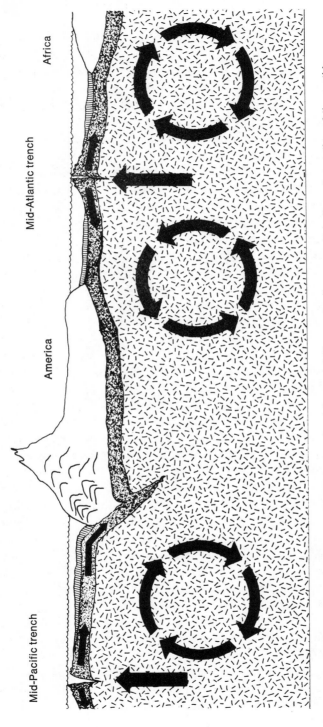

A profile view from the mid-Pacific to Africa showing the rising and falling convection currents in the outer portion of the earth's mantle that provide the motive force for the drift of the plates

ing without friction. Here, too, the collision is violent; portions of crust are shoved together and heaped up on each other like floes of pack ice that have been driven together by the movement of the waves.

But the most important mountain-building process, which takes place on the border line between crustal plates of differing density and composition, is the one geologists call orogeny: the development of mountains by deformation of the crust. The orogenic cycle begins with the forces of weathering, which transport disintegrated mineral material, in the form of pebbles, sand, and dust, from a continent to an oceanic basin at its foot. At least 5 million tons of crustal material *daily* are carried in this way to the geosynclinal trough at the foot of· the continent and piled up, layer on layer, along with the oceanic sediments. To transport such a quantity would require a freight train stretching all the way from New York City to Omaha, Nebraska, about 1,150 miles. The Colorado River alone carries—as we have seen —500,000 tons of material into the Pacific Ocean (see Figure A in the drawing on page 113).

The geosynclinal troughs, which collect thousands of feet of sediments—chiefly the continental erosion materials—are the germ cells of future mountains. As we know, the troughs are located where the convection currents of the mantle are directed downward and the oceanic crust is dipping deeper into the mantle. The lighter continental deposits—carried along in the general downward movement— ultimately reach a depth at which the denser and heavier mantle rocks are normally found. In the meantime, the deposits have solidified into sedimentary rocks, which are squeezed, compressed, and eventually melted by the surrounding mantle rocks. As they flow downward to greater depths, they undergo plastic deformation and folding. In due time a complete mountain is formed as much as 60 miles deep in the interior of the earth.

As the sinking movement continues, it increasingly disturbs the density relationships within the mantle, until a critical boundary is reached. Then the sinking movement shifts to a rising movement. Like a piece of wood that has been pulled downward by a strong whirlpool but comes shooting to the surface again because of its low density, one day the lighter, now folded continental sedimentary rocks float to the surface of the mantle in the form of a largely finished massif. The mountain that was created in the depths of the earth becomes visible on the surface and is attached to the existing mountains of the continental rim. As with an iceberg, however, only the very peak of the massif—a mountain thousands of feet high—reaches above the surface. The greater part of the massif—an almost unimaginable 230,000 feet in depth—remains buried deep in the earth's mantle, hidden from human eyes.

A mountain chain such as the Andes, for example, is never folded and elevated all at one time. Sinking, folding, and elevation proceed by stages. While one folded and elevated portion of the continental crust is again being subjected to weathering, a new geosynclinal trough forms in front of it. The trough is filled with deposits from the erosion of the elevated mountain chain (see Figure C, opposite). Subsequently these deposits, too, are folded, raised, and annexed to the existing mountains. Another geosynclinal trough will then pass through the same orogenic cycle. Depending on the width of the folded mountain mass, the geosynclinal trough will be displaced farther off the coast, and so the dipping zone for the oceanic plate will move steadily outward.

All these processes flow into one another continuously, with no sharp transitions. Mountain formation proceeds in waves, constantly involving new parts of the continental crustal area. The light, "undrownable" continents drift and fold more and more. What they lose in horizontal extension they gain in vertical extension.

The collisions and foldings of the light continental crustal regions over geological eons have, of course, led to a thickening and heightening of the continents. At the same time, the deep sea basins have grown steadily wider. The upfolding of mountains in the past 600 million years, about which geologists are especially well informed, has been subdivided into three principal periods of activity: the Caledonian, the Varistic, and the Alpine foldings.

From the basic structures and the foldings of the oldest mountain chains, it can be proved that the more recently created mountain chains have been steadily increasing in height right up to the present Cenozoic era. The reason may be that the space available for the mountain-forming geosynclinal troughs diminishes with each successive era of orogeny, causing the masses of rock to be squeezed tighter, folded more, and heaped up on one another to a greater extent as time passes. The foldings in what are for now the youngest mountains—such as the Alps—are much more striking than in older mountains.

Since these higher mountains tend to form chiefly on the edges of the continents, it would seem that the earth is condemned to an

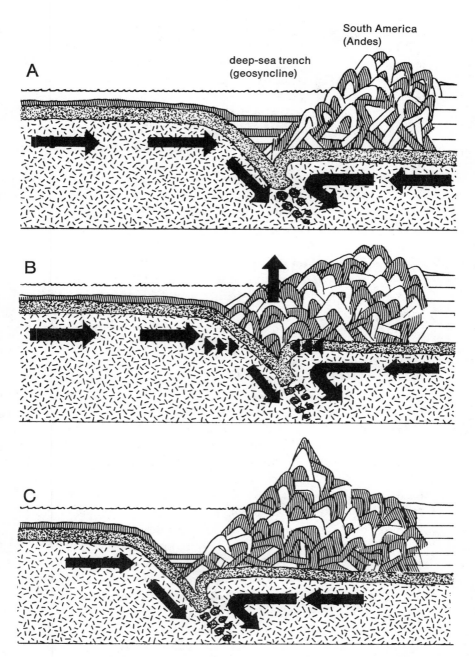

Orogeny advancing in stages in the Pacific area, as illustrated by the westward-drifting South American plate

increasing concentration of its hydrosphere in the oceans. That is, it will become more and more difficult for rain clouds from the oceans to pass over the ever higher mountains and make their way into the interior of the continents. Apparently we must expect an increase in the number of relief deserts.

The famous geologist Alfred Wegener, who was the first to conceive of a mobile crust and drifting continents, methodically worked out a model for the formation of mountains that showed them constantly increasing in height. He took the hypsometric (altitude-measuring) curve of the earth's surface as the geometrical expression of the thickening of the crust under the continents. He was the first to analyze the past and future of the earth as a consequence of the theoretical development of that curve.

Since there is something akin to a directed historical evolution of the relief surface of the continental crust, the populating of the continents by plants and animals must encounter certain limits. As we have seen, the flora and fauna had to wait 4 billion years, until they had prepared an atmosphere, before they could take their first steps onto dry land. The conquest of the land some 400 million years ago proved an extremely successful enterprise for a time; the continents were a good deal flatter then, and clouds could carry rain unhindered to every part of them. But as the continental crust was raised up by collisions and orogenies, the conditions of life on the continents deteriorated; today life has been driven out of vast regions.

The spherical shape of the earth and the amount of solar radiation received in the horse latitudes and the polar regions have imposed a fixed pattern of desert formation on the floating continents. In addition, as we know, practically any area of the continental crust can become a relief desert. The most advanced state of the desert-forming relief pattern is exemplified in the western half of North America and in huge areas of Asia. Under the dual action of the mountain-forming and crust-raising processes, the continental crust in these regions has become relief desert extending thousands of miles. We have already described some of the evidence that they were not always arid: the petrified remains of saurians and jungle trees in the heart of the American deserts. Those huge creatures lived more than 100 million years ago, when the climate was considerably wetter than it is now. Although the giant saurians are completely extinct, living remains of the jungles that once covered vast areas of North America still survive. These are the sequoias, those mammoth trees that reach heights of nearly 300 feet and have trunk diameters

of as much as 30 feet—the largest life forms on earth. They attain
ages of approximately 4,500 years (see Plate 13).

These trees, descendants of those found in the petrified forests
described in the preceding chapter, are an impressive testimonial to
that remote epoch when the biophase and the vegetative biomass on
the continents reached and passed their peak.

It is highly significant, in relation to the geological evolution we
have been discussing, that these enormous trees grow only on the
western edge of the continent, in the towering Sierra Nevada of California. Here, on the outer flank of the massive mountain barrier that
keeps rain clouds originating in the Pacific Ocean from reaching the
interior, there is so much rain that these giants of the plant world still
receive enough water to sustain their metabolism.

Other desert-forming processes are contributing toward the grad-

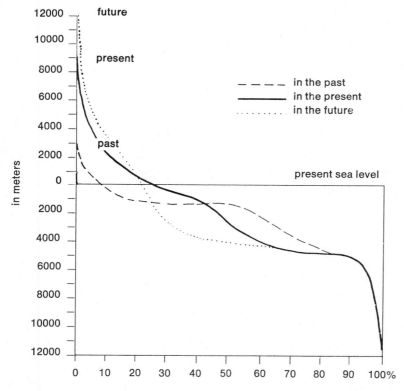

Hypsometric curve of crustal evolution. The curve shows the bimodal distribution of the heights and depths of the earth's crust above and below sea level
in the past, the present, and the future.

ual transformation of the entire surface of the continents back into extreme desert. Before we discuss that bleak prospect, let us turn to the question of how the life forms of the present geological era have contended with the steadily increasing aridity of their continental environment. In their struggle for survival, the plants and animals of the desert have evolved fantastic—sometimes unbelievable—adaptations.

# 4 / Survival Under
the Searing Sun

Of all the living organisms that exist in the desert, man is the least adjusted to arid conditions. Unlike many of the species of animals and plants that inhabit this most hostile of environments, man has evolved no significant physiological adaptations, such as water-storing organs or similar aids to survival under desert conditions. If an adult human being, naked and without drinking water or food, were placed in a shadeless and waterless desert on the morning of a hot summer day, his chances of surviving the day would be zero. By noon he would have sweated away from 4 to 6 pints of salty water, and by late afternoon, when the temperature rises to 50° C. (122° F.) in the shade, another 8 or 10 pints would have evaporated through his skin. By sundown he would have lost a total of around 15 pints of water. When the body is unable to replenish fluids lost in this way, it gives up the water stored in its fat and tissues. A considerable amount of water is also drawn from the blood. The blood becomes thicker and flows more slowly, causing a dangerous rise in body temperature. In addition, the sweat glands, which have excreted so much water to cool the body, fail from overwork and from violent sunburn. The body loses its ability to cool itself. Before evening could bring coolness to the air, the body temperature would have risen to 42° C. (107.6° F.). High fever, delirium, and circulatory failure would lead to death. Even if a person drank 8 or 10 pints of water a day, he would survive no more than five days. He would die of the effects of the burns inflicted on his body by the searing rays of the sun.

Tens of thousands of human beings have died in the desert from thirst and heat. Death was especially common during the days of the great trans-Sahara caravans of the last century, when the Arabs drove captive West African blacks by the thousands north across the Sahara. Sometimes the wells at which the caravans rested were dry or did not hold enough water for all. Sun-bleached skeletons lying near remote wells in the Sahara testify to the terrible tragedies that occurred.

To this day people regularly die of thirst in the desert. A few inexperienced tourists, in particular, succumb every year. Only a few years ago, five tourists in the Egyptian desert paid with their lives for their rashness. In spite of warnings from those who knew the area and a strict ban by the authorities, they tried to reach an oasis 125 miles away in three ordinary automobiles that were completely unfit for desert travel. By the time they had covered nearly 95 miles, all the cars had become choked with dust or were stuck in the sand, one after another. With a water supply of only a quart per day per person, instead of the requisite minimum of about 6 quarts, they set out on foot, intending to follow the tracks of their cars back to their starting point. They did not dare head for their destination, the oasis only 30 miles away, since they had no maps and did not know exactly where it was located. Even if they had hit on the idea of traveling by night and spending the day in the shade of rocky clefts, they would have had no chance of surviving over such a distance with so little drinking water per person. When, only a few days later, a desert patrol discovered the abandoned automobiles, not one of the tourists was still alive. The tracks of the cars were followed, and after only 10 miles the first body was found, mummified by the sun. One man had actually made 30 miles before he collapsed. The toughest member of the party was a black poodle that had joined the fatal expedition. Its mummy was found 50 miles from the cars, and the dog had trotted in the right direction.

In 1973 a similar tragedy occurred in an area of the northern Sahara that I know well from my own researches. During the hottest month, August, a couple set out from a large oasis for a smaller oasis only about 20 miles off. They reached it without difficulty. But in the afternoon, when they started back, they made a grave mistake. Because they had had no trouble with their car on the way and were sure they could return as quickly as they had come, they neglected to refill their water bottles. That would probably not have mattered if they had had enough gasoline. As it was, they ran out of gas about 10 miles from the starting point of the outing. While the woman remained sitting in the shade of the automobile, the man set out on foot to bring help from the oasis. Five hours later, when he returned by car with a can of gasoline, she was still sitting there. But she had perished of thirst.

These examples show how poorly equipped man is by nature to cope with desert conditions without the constant intake of drinking water. The body can be kept cool for a short while by the excretion of sweat, and the perilous loss of salt associated with perspiration can be avoided by a temporary adjustment of the kidneys and sweat glands to minimal activity. But the human body is totally unable to retard the fatal loss of water, let alone store up water as a reserve against periods of thirst.

How is it that of all living organisms man, in other respects so highly evolved, should have so few of the physiological adaptations that nature provides for life in the desert? The answer seems fairly simple. In order for there to be physiological adaptation to such extreme conditions as prevail in the desert, a very long period of evolution is required. Chance, trial, error, and success are the principal instruments of evolutionary adaptation. Only the best-adapted individuals of a species survive and transmit their abilities to their descendants. The weaker individuals, less well equipped for life in a given environment, die out. In general, each individual takes only a tiny evolutionary step. For significant adaptations, such as the development of entirely new organic functions, many generations and a time span of perhaps several million years usually are needed.

A few families of animals have been living on our earth for several hundred million years. During this long period they have had to deal with desert conditions repeatedly and have evolved new adaptations. In contrast to such families, man is a relatively new organism. There have been human beings on earth only since the beginning of the Quarternary period (the Pleistocene epoch), about 2.5 million years ago, and the transitional forms that presumably fill the gap between apes and *Homo sapiens* can be traced back about 12 million years, into the Tertiary period. The genesis of man, moreover, falls in one of the cooler, moister phases of geological history that form relatively brief intervals between the development and expansion of deserts.

The fluid dividing line between the Tertiary and the Quarternary marked not only the development of *Homo sapiens* but also a succession of ice ages. The new human species, during the longest part of its evolution, had to deal chiefly with ice-age conditions, not with a desert climate. It was only about 10,000 years ago, when the last ice age gradually ended, that the earth warmed up again and the deserts, after a brief interlude, once more began spreading. Not until the post-ice age was man, by now highly developed, forced for the first time to confront the desert's hostility to life. But 10,000 years are far too short a time for important physiological adaptations to desert life to have evolved. Nevertheless, men have lived in the desert for thousands of years. The category of desert dweller includes groups from every race of humanity, and all these peoples have devised methods for surviving in the desert. Their methods, however, are not the same kind as those that nature has evolved for plants and animals over millions of years. The human means of survival are cultural achievements.

Desert dwellers apply their intelligence to the difficulties of their environment. But human intelligence is also the final product of a long evolution. From the geological point of view, it is a matter of indifference how a given life form survives in the desert; the main thing is survival, whether by physiological adaptations such as animals have developed or by human intelligence. At first glance, intelligence appears to be an enormously useful attribute. Man, because of his intelligence, seems able to do more than merely utilize the biosphere for which he is best fitted physiologically. With his intelligence, he can simply overleap the long periods of time that were necessary for animals and plants to evolve their various adaptations. Intelligence seems to have freed man from the tedious processes of development, to have made him independent of evolution.

Man, by studying the adaptations to desert life of plants and animals, can learn to utilize the mechanisms that these organisms have developed over hundreds of millions of years. With the technological resources he has developed, he can imitate those mechanisms. The reservoirs for water, the wells, and the air conditioning that have been introduced into the desert—these are only a few examples of the technical achievements that enable people to survive in lands for which they are physiologically ill equipped. It is characteristic of the technocrats of the twentieth century to think that these successes prove how independent of nature men have become—by virtue of their intelligence. There are many who believe that man has evolved into an independently acting being who can govern his own future evolution.

Many people regard it as only a matter of time before man irrigates all the deserts of this earth and turns them into one great vegetable garden. Such ideas are wholly mistaken. They spring from a one-sided, anthropocentric philosophy. No organism in this universe, including man, will ever be able to escape the pre-established developmental constraints of its environment. Perhaps the real and profound tragedy of man lies in the fact that his aims and ideals do not coincide with the constraints of the cosmos and of the planet on which he lives.

When the earth's crust resumed large movements again, in the Mesozoic era, new mountains rose up. After a brief—geologically speaking—phase of lush development of living forms, large areas of the land became desert once more. The origin of man was still 150 million years in the future. Let us forget man for the time being and see how plants and animals dealt with the parching of the earth that has continued down to the present day. Faced with "environmental disaster," life tried to adapt.

## The Struggle of the Plants

Water is the prerequisite for every chemical reaction in a plant cell. With its aid, substances inside the plant organism are transported. Four billion years ago life came into being with the help of water, and without water no life would be possible today. Not one of the many organisms that left the sea hundreds of millions of years ago to occupy the arid land of the protocontinents has ever been able to make itself independent of water. The oldest land vertebrates, the amphibians, reptiles, and birds, still spend the first part of their developmental cycle entirely in water, like the amphibians, or in the miniature aquarium of the egg, like the reptiles and birds. The adult individuals of these classes of animals, like all other land creatures, including men, must constantly maintain a "mini-sea" in the tissues of their bodies. This "sea" is the circulating blood, which supplies the body with water as well as nourishment. In order to balance the loss of water from breathing, sweating, and excretion, the overwhelming majority of land animals must constantly seek drinking water in rivers, lakes, or ponds. Plants, of course, cannot move about to find water. Yet, of all land organisms, the immobile plants have developed the most effective method of obtaining water. Their widely ramifying root systems are able to draw the tiniest concentrations of water out of the ground, and they accumulate water much more efficiently than animals do. Plants generally thrive best where the soil is kept constantly moist, by precipitation or by irrigation.

Deserts, where little rain falls and that only at irregular intervals, are the most unfavorable biosphere for plants that can possibly be imagined. Vast areas of many deserts, especially those that receive rain only a few times in a century, are completely without vegetation. Indeed, the very idea of a desert has virtually become a symbol for a region denuded of plants. And yet some deserts, which look as hostile to life as others and which receive rain no oftener than once a year, or sometimes only every two or three years, are nevertheless a haven for a large number of plant species.

In the desert the energy of sunlight that is essential for plant photosynthesis is present in such abundance that plants have to protect themselves from it. The second prerequisite for photosynthesis, carbon dioxide, is also present—this gas is fairly evenly distributed throughout the atmosphere. But the third prerequisite, water, is not present in adequate quantities. Desert plants have therefore been compelled to make a whole series of profound adaptations to life under drought conditions. Some desert *animals* have become virtually independent of rivers and ponds; they take their water from the plants they eat. But desert plants cannot live on water accumulated by other living organisms. Except for a few special kinds of plants—the carnivorous species, which do not inhabit the desert—plants do not "eat" in the sense that animals do. They stand at the beginning of the food chain, producing their substance out of solar energy and inorganic matter. Since they are the first links in the food chain, they must be the first to obtain water on their own.

Some of the driest regions of the North American deserts are covered with dense forests, although they receive rain only once a year. Just as in the woods of rainier zones, the trees in these desert forests—trees that reach heights of 65 feet—sprout so thickly that they stand only a few yards apart. And, just as in the woods of rainier zones, woodpeckers go up and down their trunks. In one respect, however, the desert groves differ in appearance from an ordinary wood. The trees have only two or three branches and no leaves at all. Moreover, their trunks are not brown but a brilliant green. These desert groves consist of what are probably the best-known desert plants, the cacti, and the largest form of cactus, the saguaros. In addition to this giant form, there are hundreds of other species of cactus, the smallest of which grow no bigger than a thimble. Cacti are relatively new and young plant forms that probably arose during the Cenozoic era.

As the mountains rose steadily higher and as larger portions of the North American continent turned to desert, the forest became parched and many species of plants died out. But a few species of trees were able to adapt successfully to the new situation. Over the course of countless generations they gradually evolved into cacti. They learned that large leaf surfaces were unnecessary, since the reduction of cloud cover brought an increase in sunlight. In fact, they had to shed heavy foliage to prevent the evaporation of water through the leaves. The heat-sensitive leaf surfaces gradually atrophied, and the plants displaced the chlorophyll they needed for photosynthesis from their leaves into their trunks. The new plants' most important innovation was to alter the cellular structure of the trunk in such a manner that a large supply of water could be stored there for use during long periods of drought. The cacti we are familiar with today possess an extensive network of roots close to the surface, so that in the shortest possible time, during the rare but torrential desert rains, they can absorb huge amounts of water. A full-grown saguaro cactus is able to store from six to eight tons of water in its trunk. With such a reserve it can survive a dry period of up to two years. The enormous storage capacity is due to the elasticity of the trunk. Its surface is folded like an accordion, enabling the trunk to expand its volume considerably when it takes in water.

A peculiarity shared by many water-storing plants, or succulents, which include the cacti, is their spines. In many species these are so close together that the plant resembles a rolled-up hedgehog. In fertile, rainy climates, herbivorous animals can easily find water to quench their thirst; in the desert, such animals use plants not only for

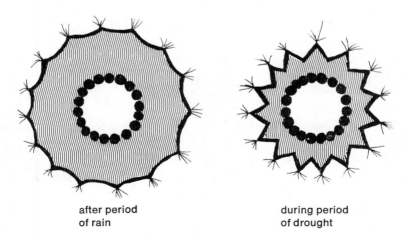

after period          during period
of rain               of drought

Cross-section through the trunk of a giant cactus

nourishment but also as a source of water. The spines of the cacti plainly serve a protective purpose, turning the plant into a redoubtable fortress.

In addition to their defensive function, the spines have other important purposes. Some cacti are covered with such a dense network of spines that the green cuticle beneath can scarcely be seen. If the network is silvery in color, as it is in some species, it reflects a large part of the sun's rays. Between the spines, which grow close together in small cushions, are countless fine jets from which the cactus can spray water. As the water evaporates it cools the plant, preventing dangerous overheating during the summer months.

Like the thermostat of a sprinkler system in the ceiling of a room, which sets the apparatus going when a fire breaks out, the cooling system of the cactus is set in motion by a natural thermostat that responds to high air temperatures. Even the placement of the jets within the pad of spines appears ingeniously calculated. The desert wind forms tiny whirlwinds between the spines, so that the water vapor is not all carried off at once and the cooling produced by evaporation can yield its maximum effect.

The most difficult problem that the newly evolving cacti had to solve was how to continue carrying on photosynthesis under the extreme climatic conditions of the desert. In order to limit the loss of water by evaporation, the cacti developed a thick, impermeable, leatherlike skin. But a skin that keeps in water vapor will also keep out the gases a plant needs for photosynthesis. Naturally, the cacti can carry out photosynthesis only in the daytime, when the sun is shining. The plants can obtain the water they need from their inner stores. But to obtain from the atmosphere the equally indispensable carbon dioxide, they would have to open the pores in their skin, as other plants do. In the desert that would be dangerous, since much water would be lost through the open pores. To solve that problem, the cacti and other succulents have developed a special trick. Reversing photosynthesis, the cactus breathes in oxygen at night (the danger of evaporating water disappears after the sun has set). The oxygen is used to burn sugar, releasing carbon dioxide. However, the cactus does not exhale the carbon dioxide, as other plants do, but stores it by chemically binding it to organic acids. By day the cactus uses the carbon dioxide for photosynthesis. If any carbon dioxide is taken in directly from the atmosphere, the cacti initiate a kind of internal circulation of it, so that photosynthesis can proceed by day without the need to absorb any more of the gas from outside.

Many species of cacti, especially the giant cactus, are dependent on the "midwifery" of another plant species. The young cactus plants are very sensitive to direct sunlight and can thrive during their first years only in the shade of other desert plants. The chief provider of shade is the mesquite tree, which is widely distributed in the deserts of North and Central America. The seeds of adult giant cacti are eaten by birds that live in the cactus groves. But the seeds, being indigestible, are voided by the birds, usually at night when they perch in the mesquite trees, and the bird droppings fall at the foot of the trees. There the young cacti that grow from the seeds receive the shade they require during the early phase of their lives.

The mesquite tree, incidentally, illustrates very clearly how the cacti could gradually have evolved from trees. The mesquite represents a kind of intermediate form between ordinary trees and the cacti. Its foliage surface has been greatly reduced; it has tiny leaves with an average diameter of no more than two millimeters. Like the cactus, it has transferred the greater part of its chlorophyll into the bark of its trunk and its numerous branches, which gives them a brilliant green color. Unlike the cactus, however, the mesquite tree has not altered its trunk so that it can store a large quantity of water. It has solved the water problem in a different fashion. Otherwise, it would be impossible for giant cacti and mesquites to share the same site, since they would interfere with each other's source of water. The cacti have their extensive networks of roots close to the surface, which absorb rain water that soaks into the ground to a depth of an inch or two at the most. The mesquite tree, on the other hand, sends its roots as much as 100 feet straight down into the ground, seeking to reach deep strata of rock that contain ground water fed by rain from distant mountains. Because of this unique form of adaptation, the tree has been able to make itself independent of local rainfall. The mesquite tree in effect possesses a natural aqueduct that can bring it water from a source a few hundred miles away. Even in years of great drought, when other desert plants wither and the shrunken cacti have consumed most of their water reserve, the mesquite tree looks green and juicy and is able to evaporate several hundred pints of water a day in order to cool itself.

The life of the mesquite tree has two quite different phases. During its early years, the young plant is dependent on local precipitation. The sapling does not seem to grow at all. But that appearance is deceptive. The tree grows downward first, until it reaches a deep water vein with its roots. Only after that, when it is no longer dependent on the rare rainfall at its site, does it develop into a tall, imposing tree.

Because of the difficult conditions of the desert, all the organisms that live there are dependent on the possession of their own territory, in which they will not tolerate rivals for food and drink. Animals can actively defend the borders of their territory against intruders. It would seem impossible for immobile plants to do so, or to prevent other specimens of their own kind from settling nearby and competing for the available water. Nevertheless, desert plants do manage to keep rivals for water from settling too close. Again, the mesquite tree provides a good example of the technique.

Like peas and beans, the mesquite tree belongs to the family of legumes. Its seeds are as hard as stone. The fresh seeds, even when planted carefully and watered well, will rarely sprout; in spite of what seem excellent growing conditions, very few of the seeds germinate. But they germinate rapidly after they have passed through the digestive tract of certain animals: the protective shell of the seed can be softened and dissolved only by the digestive juices of ruminant animals, while the seed itself remains unharmed.

One of these animals is the delicate pronghorn, which has successfully adapted to living conditions in the North American deserts. Since it wanders over vast stretches, it distributes the seeds of the mesquite widely. In this way, both the spread of the species and the vital distance between single specimens of the tree are assured. As soon as the seed is excreted by the animal, it is ready for germination, and the nutritious wet droppings that surround it provide the most favorable conditions for growth.

Other desert plants have seeds whose outer coats must be removed by mechanical friction before the seeds can germinate. The abrasion is provided by the desert wind. By the time the wind has blown the seed over stony ground a considerable distance away from the parent plant, the outer shell is worn sufficiently thin so that the seed can germinate in the presence of chance moisture.

One striking phenomenon of the North American deserts is especially noticeable from the air. Huge areas are covered by various desert shrubs that store their water in thick, fleshy leaves. But the remarkable thing is that the shrubs all stand at a regular distance apart. In their struggle to obtain as much water as they can during the rare rains, these shrubs have developed an extensive root system close to the surface. The root network of a single plant can attain a total length of more than 350 miles! Through its hundreds of millions of hair roots, one plant can absorb water over an area of nearly 250 acres. A specific amount of space is necessary for each plant's survival, and that leads it to keep a set distance from its nearest neighbor. A young plant has no chance to get started between older plants, whose developed root systems promptly absorb all the available moisture. Only when an old plant dies can a young one take over the freed space. The intervals between plants of a given population always remain the same, although the intervals vary from one desert valley to another, depending on variations in the average precipitation. From these intervals, the experienced desert scientist can calculate the precipitation in different parts of the desert. In some desert plants, the need to keep other plants at a distance is so highly developed that their roots exude poisons, in order to prevent the occupation of interspaces by specimens of their own kind or by other species.

In the desert, it is easy to see how a particular biosphere can harbor only a specific number of organisms. The examples just cited also indicate, among other things, that many desert plants practice a highly effective form of "birth control" in order to assure the survival of their own species.

In chapter 1 I spoke about the "ghost rains" of the desert, in which a large part of the sparse rainfall evaporates, before it reaches the ground, in the layer of hot air that hovers above the surface, and most of the invaluable rain water is lost. One species, the so-called hairnet plant, has learned how to supply its needs from the water vapor produced by these ghost rains. But how is such a feat possible when the rain usually evaporates at an altitude of more than 300 feet above the ground, and frequently at altitudes of more than 3,000 feet? In order to "drink" this water vapor, desert plants would have to grow some 300 feet tall and, in addition, stand on their heads, so to speak, with their leaves down and their water-absorbing roots directed upward. What is more, they would have to be able to do something utterly impossible for plants: change their location with the speed of a jet plane, since ghost rains occur with extreme irregularity and in many different places.

And yet the hairnet plant, which utilizes water vapor from ghost rains, is firmly rooted in the ground, like all other desert plants, and grows to a height of no more than a foot. Such a plant could have developed only in a place where water vapor appears regularly. A spot like that is geographically fixed not only at an exact latitude and longitude but also at a precise altitude. The required conditions are to be found in just one place, in the most extreme relief desert on this planet—Death Valley, in California. This narrow valley, about 125 miles long, is rimmed by steep, mass-wasted mountainsides towering some 10,000 feet high. Since rain clouds seldom succeed in passing over the mountains and reaching the narrow valley, and since the air between the slopes of the mountains is strongly heated by the intense solar radiation, the valley is extremely hostile to life. But on a few terracelike projections along the inner side of the mountains, about 5,000 feet above the searing bottom of the valley, the plants grow, drawing their water directly from the air around them. Their presence indicates precisely the place and the altitude at which the rare rainfalls strike the top of the layer of hot air in the valley and evaporate. The location remains the same because regularly recurring meteorological conditions make it likely that at a certain season of the year a rain cloud will float in above the desert valley. Along with this fixed possibility of rain, there is a seasonally determined upper limit to the layer of hot air.

The plants themselves actually do stand on their heads. To be sure, they possess a vigorous root by which they cling fast to the stone of the mountainside and draw the mineral part of their nourishment from the ground; but their water-absorbing roots extend in the air above the leaves, which are just off the ground. In shape the plant closely resembles the African parasol acacia, except that it is much smaller. Viewed from the side, the plant's air roots form a series of layers, like parasols one above the other. From above, it can be seen that the air roots consist of a very thick, intricate web, like a hairnet. They somewhat resemble the wire mesh of an automobile air filter or a pot cleaner. The web of roots filters the air, extracting whatever water vapor is present and supplying the plant with its vital moisture. The parasols of roots also protect the sensitive leaves that produce the plant's nourishment and keep off the desert sun's searing rays.

Of the approximately 250,000 species of flowering plants on our earth, just two or three are independent of water in liquid form. The hairnet plant, which grows only in Death Valley, is one of them.

## Living Wheels and Sandfish

It is widely believed that reptiles are largely unaffected by heat and feel all the better the warmer their surroundings are. That is simply not true. No reptile can survive a body temperature of more than 48° C. (118° F.), and most lizards will suffer heat prostration at

45° C. (112° F.). Reptiles are most comfortable at a body temperature close to the normal for humans. Being cold-blooded animals—
their bodies cannot produce any warmth of their own—they have to
absorb heat from their surroundings, and they must always be on the
lookout for places that provide the ideal temperature. In moderate
climates, or in heavily shaded tropical areas, that poses no great
difficulties. But in extremely hot deserts, where, moreover, the temperature fluctuates widely from day to night, reptiles face grave problems in regulating their bodily warmth. On the bare stone surface of
the desert, which the sun's rays can heat up to 80° C. (194° F.), no
reptile could survive for more than a few minutes.

One might wonder why reptiles have evolved no obvious physiological adaptations to life in the desert, since they have had more
than 100 million years to do so. The reason probably is that their cold-
bloodedness always forced them to adapt to the temperature of
whatever environment they were in by selecting certain modes of
behavior. To survive in the desert, they simply developed the behavioral modes further along the same lines, to the point of perfection.

Of the five orders into which living reptiles fall—tortoises, lizards,
snakes, crocodiles, and the Tuatara lizard—all except the last are
found in the desert. Astonishing as it may seem, even crocodiles live
in the desert, and, what is more, in the Sahara.

Most desert reptiles escape the terrible sun by leading a nocturnal
life. They do not leave their hiding places until sunset. During the
first half of the night, when the rocks are radiating the heat absorbed
by day, the temperature is ideal for reptiles. Later the desert surface
cools down and becomes too cold for them.

At twilight, when the reptiles come out, the desert floor is still
burning hot, especially in the summer months. In order to make as
little contact as possible with the hot ground, and with the knife-
sharp edges of the weathered stone, some desert snakes have developed a strange and complicated method of locomotion. It also serves
another purpose, in ingeniously permitting them to overcome what
would otherwise be the insuperable obstacle of steep dunes made of
loose sand.

It is very difficult to describe the complicated coordinations involved, but let us begin with the starting position. The snake's body
lies diagonal to the direction of its motion, crooked into three windings, like a double S-curve. The head points diagonally forward in
the direction in which the reptile intends to move. If, instead of
looking at the snake from above, we lie flat on the ground and watch

it from the side, looking in the direction of its motion, we can see that it touches the ground at only two points: at the first bend of the body, behind the head, and at the last bend of the body, before the tail. The head, the center of the body, and the tail are raised off the ground. The creature looks something like a bent wire. Now the snake begins to move. It raises the entire first bend off the ground and stretches the whole front part of its body, led by the head, far to one side in the direction it wishes to go. The winding central portion of the body is drawn along by the same movement, and its curve, previously in the air, now touches the ground. The result is a sinuous, flowing rolling of the body to the side. The two points on which the body rests constantly shift. The rest points look as though they were continually eddying from the head down through the winding body to the tail. In the first phase of the motion, not until the last loop of the body has run out does the tail also rise from the ground, and the snake then rests on only the one bend in the center of its body. Simultaneously, the raised front portion, with the head leaning far over, has advanced somewhat. When the snake's head and the forward wind of the body are once more resting on the ground, one sequence of movement has been completed. The snake has returned to its starting position a considerable distance farther along. The force that produces the motion comes from the continual shifting of the center of gravity of the raised, curving front part of the body; it is a kind of rolling sidewise on the windings of the body.

We can visualize the movement, in simplified form, by imagining a spiral of wire rolling across the desert floor.

This peculiar form of movement leaves a very strange track in the sand, since only the parts of the snake's body that touch the ground at a given moment make an imprint. These two points on the body not only shift from front to back but also are simultaneously drawn forward by the constant shift of the center of gravity in the raised coils, so that two separate furrows, parallel to each other and diagonal to the direction of the movement, are imprinted in the desert sand. Had such tracks never been found except as the petrified impressions of some extinct species, no one would ever have deduced that they were made by a living organism.

It is interesting that we find this highly specialized mode of locomotion in two different species of snakes descended from two separate evolutionary branches. What is more, those species live on two separate land masses, which were not connected at the time the snake species diverged on their special evolutionary tracks. One of them is

The sidewise rolling locomotion of some desert snakes

the horned viper of the Sahara; the other is the sidewinder, a rattle-snake that lives in the deserts of North America.

This convergent evolution of the method of movement in two species of snakes is a good demonstration of the theory that there can be only one optimum solution for certain difficulties of terrain. What is more, it shows that many of the technical solutions applied today by man antedate man and existed long ago as part of nature's stock of "intelligence."

One day in the desert I began to wonder why nature had not invented, for the locomotion of land animals, the highly advanta-geous, energy-saving wheel. The answer seems simple enough. The constant revolving of the wheel requires that it be completely sepa-rated from the stator, the body that is being supported and moved. This separation is either at the fixed axle or, if the wheel and axle form a unit, at the end of the axle, where the wheel is driven by gears. Such an arrangement violates the principle of the coherent organic structure of all living organisms. To have a living wheel, instead of legs, as the means of locomotion for an organism would raise the insoluble problems of supplying the wheel with blood and maintaining nerve contacts with it. In addition, how would the drive —the transmission of power from the body to the organic wheel—be achieved? That, too, is possible only without a fixed connection, by means of sliding parts and interlocking gears. Further difficulties would arise, especially for desert animals, in sealing the bearings against sand and in greasing them. For these purposes technology has created sliding packings, which, however, cannot completely pro-tect axles, bearings, and gears against constant wear. Just as humans need repair shops and stocks of spare parts for their vehicles, nature would have had to set up similar facilities for animals.

We can better appreciate the insolubility of these problems if we picture an 80-ton saurian of the Mesozoic era mounted on huge wheels instead of stocky legs. Especially for a body this large, limbs made like wheels would have been a great advantage in saving enor-mous amounts of energy that otherwise had to be used for locomo-tion. But the saurians would also have had to stop at service stations for frequent checkups and lubrication, and now and then they would surely have had to replace a worn-out wheel bearing.

Nature would appear to have resolved the matter by the round-about route of putting human intelligence to work on the problem, employing lifeless inorganic materials. But if we look more closely we see that the wheel would not, after all, have made sense for the overwhelming majority of living organisms, most of which inhabit the very uneven terrain of swamps, mountains, or forests. The great merits of the wheel would really have come to the fore only for organisms living on vast plains, such as deserts. Not without reason does man build level roadways to accommodate his artificial modes of transportation. It is, in fact, significant that the idea of living wheels first occurred to me on a vast flat plain in the heart of the Sahara.

Yet nature did employ the physical principle of the wheel in the desert long before any human ever thought of it. The winding body of the snake that rolls across the ground has the same advantages as the wheel: a minimum of frictional resistance against the ground, the saving of energy through the use of leverage, an optimal ability to overcome minor irregularities in the ground. In addition, the snake, as it moves forward in its wheellike spiral, is well able to climb in loose sand. The insoluble problems of axles and bearings do not arise. And the snake can do one thing that is impossible for a mechanical wheel. By changing the windings of its body, it can alter the diameter of the wheel and make the best possible adjustment to every irregularity of the terrain.

In the desert, where so much grinding material is present in the form of sand, the living wheel of the snake's body is in the long run far superior to any mechanical wheel because it is unburdened by bearings and axles, and so it does not need servicing. This superiority is sharply highlighted by the frequency with which wheel bearings wear out. Human intelligence was not essential for the recognition of the wheel's advantages and the practical application of them. Many millions of years before human beings appeared on earth, nature's creative intelligence had recognized and applied these principles.

The locomotion of desert snakes constitutes a good argument for the idea that man does not possess a free intelligence independent of nature. Rather, nature itself is intelligent. Man is merely the temporary end product of intelligent nature, and even the greatest human intelligence can ultimately grasp only what nature permits, nothing beyond that.

Another reptile, the sand skink of the Sahara, has solved the problem of locomotion in an entirely different but equally ingenious fashion. The horned viper of the Sahara and the sidewinder rattlesnake of the American desert both inhabit regions in which areas of sand alternate with areas of rock. The sand skink, however, lives in the great regions of pure sand. Much of the sand is driven by the wind into constant movement. As it moves, the sand behaves physically much the way water does. Like the countless molecules of water, the millions of grains of sand are in a kind of fluid condition.

The sandy surface of the desert can be compared to the wind-ruffled surface of the sea or ocean. The same giant waves are to be found, but with their motion frozen, as it were; in addition, a delicate rippling, such as a breeze produces on the surface of water, stirs the surface of the desert sand.

In this "fluid" environment of the sandy desert, animals encounter great problems of locomotion. Moreover, there are vast regions where underground burrows, in which the animals could hide by day from the deadly rays of the sun, are impossible to construct because there are no roots of grasses to hold the sand.

The organisms of the sea have solved their problem of movement through the fluid, unstable element that surrounds them by swimming. Most of these organisms live below the surface, and the ones we are most familiar with are the fish. But how, exactly, does a fish swim? The truth is that the subject has not been adequately studied to this day, and what has been learned is so complicated that it is hard to summarize. We will confine our discussion to those elements of swimming that are necessary to an understanding of the locomotion of certain desert animals.

In order for a fish to move at all, it must first press the water aside. Most fish do this by snakelike, sidewise movements of the body and the tail fins. The counterpressure of the water causes it to flow along the fish's tapered body toward the tail, closing in behind the fish and constantly pressing it forward.

As far as their method of locomotion is concerned, fish could adapt very successfully to the sandy sea of the desert. In fact, one creature that lives beneath the surface of the huge sand dunes of the Sahara is called the *poisson de sable,* or sandfish. Zoologically, however, it is unrelated to real fish, but belongs to the family of smooth-skinned lizards and is therefore a reptile. Our name for it is the sand skink. It has a sharp, wedge-shaped nose, and its whole body is covered with slippery scales, which keep frictional resistance to a minimum as it moves by winding its body through the sand an inch or so beneath the surface. The extreme smoothness of the scales, which gives the creature the feel of being coated with highly polished varnish, promotes the flow of the sand from head to tail. The little lizard is extremely difficult to hold on to. Its limbs have spread and flattened to such an extent that they seem more like fins than legs. The sand skink feeds on insects that also live in the sand, and the sand is sufficiently permeable to air that the lizard can breathe. "Swimming" in the sand is not only a highly effective and energy-saving method of locomotion; it also protects the sand skink from predatory birds and from the rays of the sun.

It is as essential for warm-blooded animals as it is for reptiles to maintain their body temperature within certain limits. But these limits are much narrower than for reptiles. An excessive fluctuation up or down—fever or chills—would have fatal consequences. Warm-blooded animals, too, must effect a constant exchange of heat with their environment. Mammals and birds, which are descended from the reptiles, have an advantage over them in being able to control their own body temperature. They were the first animals in the long evolutionary history of the earth to attain independence from external temperatures, since they can keep their body temperature constant by producing their own warmth or radiating heat to their surroundings. A kind of built-in thermostat permits them to maintain just the right body temperature, whether their environment is warm or cold, whether they are engaged in strenuous activity or are at rest. Mammals and birds have developed a number of characteristics for this purpose. Mammals shield themselves from excessive warmth or cold by their hairy pelts. Birds achieve the same end with their feathers. Men, who have lost the greater part of their body hair, rely on clothing to insulate themselves from the fluctuating temperature of the environment.

The ability to keep body temperature constant has proved highly beneficial to birds and mammals. It enables them to occupy almost every biosphere, from the Arctic to the tropics. In the extremely hot environment of the desert, however, they encounter great problems in balancing the heat economy of their bodies. Some of the modes they have adopted to escape being broiled to death in the desert are almost beyond imagination.

## Birds That Build Air Conditioners

Among the more highly evolved classes of vertebrates, birds are probably the most successful, in terms of total numbers and of numbers of species, in adjusting to the desert. At first glance their greatest asset would appear to be their ability to fly. If the desert ceased to supply them with food and water, the birds could move on; also, they need not settle in the desert until, after a time of ample precipitation, living conditions have become especially favorable. And yet that is not the case with the overwhelming majority of desert birds. In order to survive in the desert even for short periods, birds have had to develop such extreme and special adaptations that they can no longer exist in other biospheres. Once adjusted to the desert, most birds are condemned to spend their lives there, even through the summer and in years with no precipitation.

However, deserts are vast regions, with extremely varied conditions inside their borders, and the birds' ability to fly does enable them to select the best places. The result is that in any sector of the desert where rain has happened to fall a large number of species and individual birds will be found gathered together. The lush vegetation that springs up quickly after a rain lures the seed-eating birds and insects first. The insects, in turn, lure the insect-eating birds. And the same area a year later will often be a parched landscape without a trace of life. In 1968 and 1969 I was able to observe an especially dramatic population change in a single species of bird, the desert warbler (*Sylvia nana*), which is one of the smallest species in the Sahara.

The spring of 1968 was unusually rainy, and between 300 and 400 pairs of warblers nested in an area of about 225 square miles. In the spring of the following year, during which no rain fell, the warblers in the same area were down to 4 pairs—a reduction of 99 per cent.

A quality typical of desert-dwelling birds is that they use flight only to find a more favorable environment. When they have found it, they set about nesting and raising their young. During this period, which lasts from two to three months, some species almost never are seen flying.

One especially good example of that habit is the cream-colored courser (*Cursorius cursor*). This beautiful bird is the long-distance runner among Saharan birds. I have come upon the bird countless times in my nine long trips through the Sahara, but only five times have I been able to observe it flying. Each time, the flying birds were part of small flocks hunting for patches of vegetation. The courser is so swift and graceful a runner that it has been called the gazelle of birds.

The running ability of the courser and several other species of birds represents a highly interesting adaptation to desert life. Running uses many fewer muscles than flying, and that makes it a much less strenuous method of movement. Also, the courser feeds on insects that live hidden under stones and sand, and it could not possibly discover its food from the air. Observing this bird as it hunts for food is a unique experience. With its long legs, the courser rapidly covers a distance of 30 to 60 feet. Abruptly it stands still, turns its "ear"—an auditory orifice capable of very fine discrimination—toward the ground, and listens carefully to whatever might be there. If it hears something, it at once darts its sickle-shaped beak into the ground.

Perhaps the best known of desert birds is not the ostrich—which is a native of the steppes and savannas—but the greater roadrunner (*Geococcyx californianus*) of the American deserts. This bird, which spends almost its entire life running, attains a ground speed of 25 miles an hour. If that speed is considered in relation to the bird's size—it is about as big as a pheasant—it is the fastest running bird in the world.

Birds, like their ancestors the reptiles, cannot cool the surface of their bodies by sweating, as mammals do. They depend on the far less effective method of evaporating water from their mucous membranes. That is why, during times of great heat, birds can be seen panting with their beaks wide open. The fact that they do not sweat, and thus are not susceptible to water loss, to some extent helps their survival in the desert; on the other hand, the inefficiency of their cooling mechanism complicates their life in the desert heat. Desert birds have been forced to discover other ways to reduce body temperature.

Gliding birds, such as the peregrine falcon, the lanner falcon, and the brown-necked raven, escape the worst heat of the day by letting the rising currents of warm air carry them to great heights. Especially during the hot summer months, these birds spend many hours of the day gliding in the cooler air more than 3,000 feet above the surface of the desert. Other species of birds escape the deadly heat in a different manner. One small Saharan bird called the white-crowned black wheatear lays its eggs in the cool abandoned burrows of small rodents a few feet beneath the hot surface of the desert.

In North America, the burrowing owl likewise takes over the disused burrows of the other desert dwellers. A factor in the owl's favor is that some animals, such as the desert fox, dig as many as eight different dens, or reserve burrows. However, there is always the danger that the burrowing owls, especially the young chicks, may be eaten in their cool underground nests if the real "landlord" happens to pay a visit to its property. As protection against this danger, the owls have developed a highly interesting mode of defensive behavior, resembling that of the dreaded poisonous rattlesnakes, which also live in unoccupied burrows.

Snakes living in hot regions that offer little shade must lead largely nocturnal lives in order to escape the heat of the day. But their principal enemies—foxes and coyotes—are also of nocturnal habit, and the snakes need some means of frightening them off. Optical signals would be pointless at night. But the nocturnal mammals of the desert have an excellent sense of hearing, and so the rattlesnakes have developed an acoustic warning signal. This is a rattle at the end of the tail consisting of several horn rings linked together, with enough space between them to produce a noisy rattle when the tail moves. The horn rings are the vestiges of skin sloughings; rattlesnakes differ from all other snakes and lizards in not casting the skin of their tails. The scales that remain gradually harden into thick rings of horn, and at each casting of the skin a new ring is added. If a rattlesnake is threatened in the open, or if an enemy attempts to penetrate into its lair, it raises its tail, and the movement of the horn rings produces a rattling sound. The intruder understands from this that a snake's poison fangs await, and it retreats in fear. Since the bite of most species of rattlesnake is absolutely fatal, it is something of a mystery how any animals could have acquired the instinctive fear reaction and transmitted it to their descendants. Perhaps at the beginning of their evolution rattlesnakes possessed a relatively feeble venom, which made snakebite merely a painful experience, rather than a fatal one. But the whole question of the transmission of reactions that assure survival is a very sticky one in evolutionary theory.

Let us return to the burrowing owls. When baby owls in an underground nest are threatened by an enemy, they produce a noise that deceptively resembles the warning of the rattlesnake. This "rattling" so frightens the intruder that it immediately retreats from the dark entrance of the burrow, without taking a good look at the inhabitants. The young owls produce their defensive sound by rapidly clacking their beaks and fluttering their wings. This successful defensive reaction is not "copied" from the rattlesnake; it probably springs from a powerful fear reaction, in much the same way as human beings in a state of terror find their teeth chattering or their knees knocking. At some point in the course of their evolution, this behavior of burrowing owl chicks developed, and, since it effectively drove off the enemy, the young birds survived and transmitted the behavior to their descendants.

. . .

Many species of birds are equipped neither for soaring flight nor for withdrawing from the sun's rays into underground burrows. How do they survive in the desert? There is almost no limit to their ingenuity.

One characteristic of most species of desert birds is the harmony between the color of their plumage and that of the desert floor. They blend so perfectly into the color of the desert in which they live that only a practiced eye can detect their presence. In an area where the average person may not observe a single bird, a practiced ornithologist will see ten or fifteen different species. Such camouflaging is absolutely essential in the wide-open terrain of the desert, where there is no cover for the birds to hide in when enemies approach. They press themselves flat against the ground and trust that their resemblance to the stone or sand will make them invisible. Time and again in the desert I have had the same experience: birds running along the ground, which I have been watching through my binoculars, have vanished as if the earth had swallowed them the moment they stopped moving. Even a falcon hunting from the air, so far as I have been able to observe, apparently cannot spot birds that are crouching motionless on the ground. Such protective coloration has undoubtedly developed largely by a process of selection. That is, the enemies of the ground-dwelling desert birds—raptorial species like the falcons—were more likely to catch the individuals of a species that differed most strikingly from the coloration of their surroundings because those birds were the most noticeable. Birds whose coloring harmonized with the surface of the desert would not be detected and would survive and reproduce, passing on to their descendants the successful coloration.

But in many cases the process of selection seems an insufficient explanation for the origin of desert coloration. For example, almost all the mammals of the desert, even though they are entirely nocturnal and hide in their burrows by day, harmonize in color with their surroundings. Yet the protective coloration can be of little use to them, since colors presumably cannot be distinguished at night. The scientific literature makes frequent reference to this contradiction. The one explanation for the phenomenon offered so far is that heat and aridity affect pigmentation, producing a pale, neutral coloration. That suggestion seems inadequate to me and, in many respects, biologically wrong. It fails to explain why, for instance, several different colorations occur in a mammalian species of nocturnal habit.

A particularly good case of such diversity is the wholly nocturnal jerboa, or desert rat, which in the Sahara lives chiefly in the regions of golden yellow sand dunes. Its fur is the same golden yellow color as the sand. But several groups of the species live in the gray and blackish wastelands of stone that surround the regions of sand dunes. These jerboas have fur of a gray coloration; once again they have adapted exactly to the color of their surroundings. In other words, the jerboa diverges from the usual neutral desert coloration. Their differences in color adaptation seem to have arisen, like the protective coloration of birds, as the result of optical selection, and this could have been set in motion only by their enemies, the desert foxes and owls. But how could that happen with nocturnal animals?

Anyone who has had the good fortune to experience a night of full moon in the desert will confirm that moonlight floods the landscape with a brightness that must be seen to be believed. One can read a newspaper without difficulty, and such landmarks as buttes and monuments can be seen clearly at a distance of 10 to 20 miles. In the sandy desert, especially, nights of full moon are so bright that a driver can effortlessly follow a track even while wearing sunglasses. The brightness at the time of the half-moon corresponds roughly to the brightest nights of the full moon in temperate zones. The reason is that in the desert the moonlight is not blocked by strata of clouds or filtered by atmospheric moisture. Also, desert sand frequently consists of tiny crystals of quartz, which reflect the moonlight in their myriad lenses.

The bright moonlit nights offer an excellent opportunity to observe the shy mammals of the desert about whose biology and behavior little is known.

To find out something about the origin of camouflage colors in the nocturnal mammals, I have made a number of experiments with jerboas on nights of full moon. For instance, I caught several golden yellow specimens in the sandy desert and several grayish ones in the adjacent stony wasteland. I released the sand-colored animals on the edge of the rocky area, and their bright fur stood out distinctly against the much darker ground. It was also interesting to see that the animals promptly rushed back to the sand dunes where I had found them. The gray jerboas likewise fled toward the rocks that were their normal ecological niche.

In light of my experiments, and because the nocturnal mammals, both the hunters and their prey, have excellent night sight, there is reason to believe that the protective coloration of these desert animals, too, arose as a result of the normal processes of natural selection.

Some of the diurnally active birds of the desert are found only in very limited areas, apparently restricted there by their special coloration. One of the smallest birds of the Sahara, the sand-colored desert warbler, nests nowhere but in the sandy part of the desert. Other bird species show distinctive color variations, as the jerboas do. The desert lark (*Ammomanes deserti*) is widely distributed in the Sahara; it is perhaps the commonest bird in that great desert. More than twenty different color variations of the species have been described. It might almost be said that the bird has as many different colors as do the rock formations of the Sahara. I have carefully studied this phenomenon in a special sampling area about 6 miles wide and more than 35 miles long. Within those 210 square miles I saw an amazing number of varieties of desert lark. The dominant feature of the area was a fine, gray brown rock. Interspersed with it were patches of coarse black manganese rock and reddish brown sandstone. Then there were other spots, some measuring only 1 or 2 square miles, whose rocks were covered with a thin layer of yellow drifted sand. A different-colored mini-population of desert larks lived in each one of these adjacent neighborhoods. There were yellow, tan, gray brown, blackish, and reddish pairs of larks. Each pair's plumage harmonized in color with the particular part of the desert floor it inhabited. An extremely effective process of fine selection must have been at work here, leading to the assumption that many enemies were present. In fact, that was not the case. The entire sample area boasted only a single pair of peregrine falcons, and as far as I could observe they did not pay any attention to the small desert larks. They hunted chiefly the larger species of birds and were exerting no selection pressure on the larks.

Eventually I discovered the answer to the mystery in the behavior of the desert larks. The birds had a habit of taking dust baths with striking frequency. In the process, their plumage took on the coloration of whatever ground they happened to be occupying. The desert larks, then, had not adapted genetically to the different colors of the desert as the result of direct selection pressure exerted by their enemies. Their adaptation was behavioral. But selection pressure operated indirectly, since the birds that survived were those that took the most frequent dust or sand baths—for they were the best camouflaged.

A particularly interesting problem arises in connection with the protective coloration of these birds. Individual members of the species are as much camouflaged from each other as from their enemies. Since the density of the bird population is already low, due to the extremely poor living conditions of the desert, the effect of the low density is reinforced. In my ornithological sample areas at the northern rim of the Sahara, I generally found only one pair of a particular species of bird nesting in an area of about 14 square miles. That is an especially sparse population in view of the fact that my sample areas offered some of the best living conditions possible in a desert. In other parts of the Sahara not a single bird can be found in an area approximately 185 miles square—nor, for that matter, any other highly developed organism. The huge distances, in combination with the protective coloration, make contact difficult within a given species of bird, and hamper mating.

In overcoming the tremendous distances, desert birds have developed a number of interesting adjustments. They employ both visual and acoustic signals. The visual signals are hidden carefully in the plumage for almost the entire year. Since it is perilous for the birds to attract attention, the signals are brought forth only during the brief period of the mating search.

The cock of the rather turkeylike Houhara bustard (*Clamydotis undulata*) represents a particularly good example of such a signal. Normally, this bird perfectly matches the rocky part of the desert that is its habitat. But during the mating period it stands atop some elevation in its terrain and puts on a display of dual coloration. To attract attention, it virtually turns its entire plumage inside out. It raises its tail feathers and twists its wings to expose the white underside of the feathers. The feathers of the long neck are also puffed out, and a dramatic color signal is produced by their white underside in conjunction with two black streaks on the plumage of the neck, which normally form part of the protective coloration. The bird, completely changed in appearance, can be spotted from a distance of 1 or 1½ miles in the open terrain of the desert. To appreciate the far-reaching effect, consider the range over which the mating signal can be seen. Assuming that the bustard hen can detect the signal at a distance of a mile and a quarter, this becomes a circle with a diameter of 2½ miles, and the total area over which the bustard cock is signaling to the hen "Here I am" is more than 4½ square miles.

Small species have an even greater problem finding mates in the boundless expanses of the desert. The bifasciated lark (*Alaemon alaudipes*) is a typical ground runner and is seldom observed flying. The bird has developed a highly effective visual signal in the form of a striking black-and-white marking on the upper coverts, which can be seen only when the wings are spread. With the wings folded, the mark is so changed that the bird has complete protective coloration. In order to display its color signal with the fullest effectiveness, the male has developed a curious kind of mating flight. It rises very rapidly, with flapping wings, to a height of about 30 feet, and then, with wings outspread, it dives vertically toward the ground. The marking on the wings becomes visible from a great distance. During this mating flight the bifasciated lark runs a considerable risk, for it may attract hungry falcons as well as the females of its species. The striking markings on the wings, along with the need to hunt for food, are probably reason enough for the bird to keep to the ground except in the mating period.

Many desert birds have evolved interesting acoustic signals that reinforce the effectiveness of their visual ones. If you tramp through the desert in the spring of the year, you will hear, in many places, a unique acoustic phenomenon that is quite unmistakable but difficult to describe. The air will be filled with thin, high-pitched sounds not normally heard in nature, so that it is difficult to fit them into one's ordinary acoustic experience. The pitch is so high that it approaches the limit of human perception. The sounds are reminiscent of the measuring signals one occasionally hears over the radio, or of the high-pitched note that comes from a loudspeaker when the microphone has been overloaded.

The sound is the mating song of the male bifasciated lark. Its singing bears no resemblance to the song of other birds. Complicated stanzas are not needed in the desert; they serve for areas with many species of birds that must be distinguished from one another. The desert has so few species that identification problems rarely arise. The "song" of the bifasciated lark has as its primary purpose to be heard over a great distance. It is almost always linked with the visual signal of the mating flight. When the male rises in the air, the song begins with a rapidly climbing *dyee, dyeee, dyeeee,* until, as the bird dives, the highest pitch is reached and only a long-drawn-out *eeee* can be heard. The sound is audible over a few miles' distance, by far exceeding the range of the visual signal. Moreover, the high pitch is especially well suited to penetrate many strata of heated air.

Human beings employ a similar system of signals. If we do not succeed in attracting someone's attention by waving, we often utter a shrill whistle.

One of the commonest and yet most unusual species of bird inhabiting the Sahara is the white-crowned black wheatear. It seems to refute everything we have learned up to now about protective coloration in desert dwellers. The wheatear, which inhabits rocky and mountainous regions with almost no vegetation, is probably the most striking creature in the desert. It is about as big as a chaffinch and has intensely black plumage. But the feathers of its back, at the rump and in the region around the base of the tail, where the coccygeal gland is located, are brilliantly white, as are most of the tail feathers. The top of the head also is white, probably to protect the brain from becoming overheated.

The first time I encountered this bird in the Sahara, many years ago, I was baffled by its striking appearance. The bird was perilously conspicuous, I thought. But I soon realized that what had been written about it—that the white-crowned black wheatear is the commonest species of desert bird—was not true. Strictly speaking, the species is neither more common nor more rare than, say, the larks of the Sahara. Its "commonness" is due solely to its conspicuousness; in contrast to most of the other birds, with their perfect camouflage, it calls attention to itself and therefore is seldom overlooked by travelers in the desert. What is more, living as it does in remote areas that are difficult of access, the wheatear has never encountered human

White-crowned black wheatear

enemies, and so most of these birds show no fear of man. It is often possible to come within a step of a wheatear without its flying away.

But few desert travelers have had occasion to observe the bird's behavior when its real enemy, a falcon, approaches. Then the extremely alert bird finds safety in an optical trick. As soon as it catches sight of a falcon, it flees to one among its millions of hiding places: the innumerable black shadows cast by the cliffs and rocks of its rugged habitat. Since the eye adjusts itself to the glaring brightness of the desert, the shadows under overhanging cliffs appear coal black. The instant the white-crowned black wheatear crouches in the shade, it looks to an observer as if the bird had vanished behind a black curtain. Evidently the falcon is deceived in the same way. The round white dot of the head and the narrow white streak of the tail look like the small bright markings left on the surface of a cliff by sunlight penetrating through small, weatherworn holes and crevices. Within its particular habitat, the ostentatious appearance of this bird turns out to be an extremely subtle form of protective coloration. For birds living in the level, shadowless parts of the desert, such coloration would, of course, be useless.

On the other hand, when the white-crowned black wheatear perches on top of a boulder or at the edge of a cliff, it has no difficulty advertising its presence to a possible mate. In the sunlight, its chiaroscuro camouflage becomes a signal visible at a great distance.

One problem that confronts all desert birds is the integrity of their territory. Where living conditions are so poor, the birds must take care that no other members of their own species—except during mating and nesting time—enter their territory. Other species can be tolerated because each specializes in its own type of food, and the species do not compete with one another. The white-crowned black wheatear marks off its territory by flying along the boundary lines— "beating the bounds," so to speak—once a day. This it does in a very leisurely fashion, repeatedly perching for a while on a prominent projecting cliff and looking around for neighbors that might want to make a territorial claim. Its ostentatious coloration is an aid to the wheatear in marking out the boundaries of its "property." The bird acts like a kind of flying fence.

It was on the cool winter day of February 24, 1968, that I discovered this bird's most unusual habit. I was watching a white-crowned black wheatear struggling to move a stone that was enormous for a creature its size. That angular stone measured close to an inch square and was about a quarter of an inch thick. The wheatear had taken the stone—which must have weighed nearly as much as the bird itself—in its beak and, half flying, half hopping along the ground, had vanished behind a projecting cliff. After a while it returned to its starting place. Peering through my binoculars, I saw that the ground was littered with thousands of pebbles, which had probably been washed down by a heavy shower of rain. The bird carefully selected another pebble and moved that one with equal difficulty. It came back to its miniature quarry a third time. Now I followed the toiling bird, for I wanted to learn the meaning of this curious procedure.

The bird briefly vanished from my sight, but I discovered it again on a small projection in the shadow of an overhanging cliff. It was engaged in adding the new pebble to those it had already accumulated on this projection, which was about 8 inches across. I could not fathom the meaning of the pebble collecting, so I watched the spot for weeks. Often the bird would not approach its building site for days. On other days it would spend many hours laboriously tugging and carrying pebble after pebble. It carefully piled one pebble atop another, gradually creating, over a period of some weeks, a pyramid of pebbles about 6 inches high. One side of the pyramid leaned against the massive rock of the cliff, and the pile was arranged so that at the top there was a bowl-shaped depression about 5 inches in diameter. This hollow was lined with stubby bits of wood. Now there could be no doubt that the wheatear was building a nest. The base of it, according to my count, consisted of some 360 small pebbles. It was a truly incredible labor for a small bird, especially when each pebble weighed at least half as much as the bird itself (see Plate 23).

With the aid of a thermometer, I discovered the advantages of this unique form of nest building after the bird began incubating its eggs in the spring.

Because of its unusual protective coloration, the bird cannot nest in the parched shrubbery, where it would stand out too clearly. Besides, the bushes would not offer enough protection from the sun's rays. The bird therefore builds its nest in the permanent shade of an overhanging cliff.

23. The nesting pyramid of a wheatear, built out of several hundred pebbles, in the shade of a rock. The pebbles allow good ventilation and diffuse the heat; the actual nest, made of plant materials, is in a hollow at the peak of the pyramid.

But solid rock is a good conductor of heat. The cliff heats up rapidly in the spring, when the intense solar radiation begins striking its surface. Even in a shaded area the temperature rises high enough to harm sensitive eggs.

However, the pyramid of pebbles that the bird builds is a poor conductor of heat because of all the air spaces between the pebbles. Also, the wind that blows constantly in the desert provides cooling ventilation at the base of the nest. The most important cooling effect, though, is due to the nature of the pebbles themselves. They are chiefly of porous sandstone.

It may come as a surprise to learn that dew falls in the desert. To be sure, there is little moisture in the air, but the rapid cooling of the ground at night causes traces of atmospheric moisture to be deposited on rocks as a thin film of dew. The porous sandstone under the nest absorbs this dew like a sponge. Elsewhere, the almost invisible film of dew evaporates well before sunrise in the growing heat of the air. But the sandstone pebbles under the nest release their stored moisture slowly during the day. The resultant evaporation cools the nesting pyramid.

Of course, the bird is not able to distinguish types of stone as if it were a geologist and could invariably choose the right building material. It picks out the pebbles that are lightest and easiest to transport, and these also happen to be the most porous ones.

The white-crowned black wheatear builds a new stone pyramid every year. It performs this strenuous work during the cool winter months, but the nest is not occupied until brooding time, in April. Essentially, with this unique form of nest building the bird is setting up its own air conditioner in the desert. It is creating a microclimate within its hot environment, thus providing its eggs and chicks with a good chance to survive.

## The Secret of the Rock Doves

As the name *rock dove* suggests, this species of pigeon inhabits rocky, mountainous regions. The birds usually build their nests on steep, inaccessible cliff walls, either on small projections or in crannies.

The rock dove is a beautiful bird with softly gleaming, blue green plumage. I was quite familiar with it, for it is common in the mountains that border the northern rim of the Sahara. With binoculars I had often made out rock dove nests at dizzying heights.

Also, for several years I had regularly observed the bird, which is probably the ancestor of our common domestic pigeon, in areas far beyond its usual mountainous range: in the low-lying hot plains of the desert itself. During the cooler months, great flocks of the birds, often numbering several hundred, gathered in the Sahara proper.

It was an aesthetic pleasure to watch the elegant maneuvers of the flocks as they flew, in a gleaming pigeon-blue cloud, over the sun-parched brown or black surface of the desert. Their striking color, which in this landscape made them visible from a great distance to their mortal foes, the peregrine falcon and the lanner falcon, showed that they had not adapted to this kind of terrain and were not at home here. I concluded that they moved into the desert only during the winter months, when it became too cold for them in the high mountains and when the seeds of desert plants could still be found in the wadis. At least, that was my initial explanation for the presence of rock doves in the desert.

In time, however, I also observed them there during the hot months of spring and early summer. At those seasons they no longer gathered in great swarms, but were to be seen only individually or in pairs. Their secretive, cautious behavior suggested that they were nesting. But where? The only possible brooding places seemed to be the ground of the endless rocky wasteland or the cliffs of isolated buttes or monuments.

The surface of the rocky wasteland could be promptly eliminated. Even in the spring months, the temperature of the rock frequently rose as high as 50° C. (122° F.). Moreover, the doves, whose coloration did not at all harmonize with that of the desert floor, would easily have been spotted by hunting falcons at a distance of nearly a mile and would have proved an easy prey. Nor did the stone pillars of monuments, which I carefully investigated, appear to be the nesting places. The sunny side of these crags, being without shade, was too hot, and on the shady side the few projections in the otherwise smooth rock were already being used as nesting places—and, what is more, by the falcons. It seemed impossible for there to be any cool, safe nesting spot for rock doves in this hot, shadeless desert—all the less so since the doves are seed eaters and are absolutely dependent on having drinking water several times a day. Was it conceivable that there were cool underground rivers flowing deep beneath the surface of the desert, and that the doves brooded their eggs on riverbanks? Or that the doves built nests that floated on the thermal upcurrents at a great height, where they could be kept cool by the rain that evaporated before it reached the ground?

Both ideas appear equally absurd. Yet one of those fantastic nesting places actually exists. I discovered it quite by chance. One day, as I stepped to the brink of a foggara shaft with a pail tied to a long rope, intending to draw water, I thought I briefly glimpsed a rock dove that whirred up from the black abyss and was immediately swallowed up again in the darkness of the deep shaft. Brief though the apparition was, I was sure I had not been mistaken. Apparently the rock dove had been about to leave the shaft, but when it saw me blocking the exit it dropped back down.

Was it really possible that rock doves nested in the underground foggara system?

But first I must explain what a foggara is. That is the name given to underground irrigation channels, man-made, that are beyond doubt one of the most interesting schemes ever conceived by desert dwellers in their struggle for the most vital necessity of life, water. Foggaras were built in the deserts of Persia thousands of years ago, and from there this method of irrigation reached the Sahara.

The method was based on the observation that certain rock layers rising above the surface of the desert—those layers that have been tilted by the movement of the earth's crust—continue below the surface, where they often plunge deeper and deeper. Men also learned that some of the rock strata, especially those consisting of porous sandstone, could store great amounts of water. An easy way to see how rocks can store water is to let a brick stand in water. From the air bubbles that rise to the surface and from the marked increase in the weight of the brick, it is quickly apparent that it soaks up water like a sponge.

I have already described how rainfall occurs in the desert, usually in the form of highly infrequent cloudbursts. During these rains a great deal of water follows the tilted strata of porous rock down into the earth. Deep beneath the surface, stored in the innumerable tiny pores in the rock, the water is safe from evaporation. Desert dwellers, who have grasped these relationships by long observation and reflection, tap the water-bearing strata by laying long underground canals at depths of 30 to 100 feet below the surface of the desert. The canals, which have a slight pitch and are often as much as 30 miles long, conduct the water to distant oases, where it is used for irrigating the fields.

In order to locate a water-bearing stratum, the inhabitants of the oasis first dig a number of deep, probing shafts. When one of the shafts hits a water vein, more shafts are dug, about 100 feet apart, in a straight line all the way to the oasis—dozens of shafts, or, if the distance is great, hundreds. At the bottom of the shafts the actual irrigation canal is dug, connecting the whole line of them, as shown in the diagram. The foggara system of an oasis can extend a few hundred miles. All the foggaras were dug by hand, without the aid of machinery—an almost inconceivable achievement to us who belong to an industrialized society.

Now back to the rock doves. In order to prove my conjecture that they were nesting deep beneath the surface of the ground in the man-built foggaras, I had to discover a nest. The obvious way would have been to descend by rope or rope ladder one of the 65-foot-deep shafts leading to the irrigation canal. But the walls of these particular shafts looked as if they were on the point of caving in, and that seemed like too risky an undertaking. Since I had no desire to be buried alive, I gave thought to alternative methods and hit on the idea of lowering into the shaft a camera that could be triggered from a distance. Of course, that was easier to plan than to do. How, for example, was I going to keep the camera from plunging into the water at the bottom? I would have to find a way to make it hang at the right height and in the right position. My first step was to measure the exact distance from the surface down to the water level. For this purpose I marked a long rope with paint, placing my marks at 1-meter intervals. At the end of the rope I fastened a large, empty water canister. Then I laid a long wooden pole across the top of the shaft, to serve as the ground level for my measurement, and lowered the canister from the pole. Both by sound and by the slackening of the rope, I was able to determine when the canister hit the surface of the water, and that told me exactly how many meters down the water was. At that point I marked my rope. But something very interesting happened when I lowered the canister. Rock doves flew up from the foggara shafts to my left and right. The canister had evidently frightened them out of their underground hiding place. That seemed to confirm what until then had been merely a guess. Still, only a photo would count as final proof. I wanted one showing either a brooding bird or a nest in which the eggs could be plainly seen.

I mounted an electrically operated camera with flash equipment on a board, at each end of which a rope was fastened. (Two ropes were necessary to control the position of the board.) I assumed that the pipe-shaped irrigation canal had a diameter of about 1½ meters, and so I lowered my device to a point I estimated to be above the surface of the water. By means of the two ropes, I swung the board around so that the lens and flash bulb of the mounted camera pointed in the direction in which the canal ran. That was easy to see, since it was marked by a long string of shaft openings visible on the surface.

Via the electric cable, I triggered the camera and flash bulb. Then I turned the board around and took another picture in the opposite

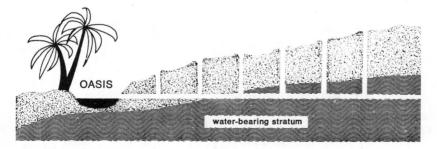

Diagram of foggara irrigation

direction. I developed the film the same day—with disappointing results. I could plainly make out a short stretch of illuminated rocky tunnel with water flowing at the bottom, and there were many broken pieces of rock lying in the water. But neither negative showed any sign of a bird or any indication of a nest.

Over the next few days I lowered the camera into other shafts and took many more photos. It was the thirty-sixth photo that at last provided the proof I was seeking. The picture plainly showed a large nest on a fallen rock, with two black eggs in the nest. Of course, the eggs looked black only on the negative; in reality they were a brilliant white.

Deep beneath the surface, inside the cool foggara system, the rock doves are shielded from the deadly rays of the sun and, in addition, have an ample supply of drinking water. They are also shielded from their foes, the falcons, while they are brooding and raising their young. The older birds leave their dark home during the nesting period only to seek food. When the chicks are fledged, they probably train their flying muscles for days along the underground irrigation canals before they are able to undertake the difficult flight straight up one of the narrow shafts into the light of the world.

This newly discovered breeding site of one species of bird provides further evidence of the remarkable adaptations that living organisms are capable of when they must find a means of surviving in an environment hostile to life.

## Sand Grouse—Masters of Survival

What are undoubtedly the most fascinating and clever adaptations to life in the desert are those that have been developed by several species of birds belonging to the sand grouse family. The name is somewhat misleading, for it is unclear just how these birds ought to be classified. They look somewhat like pigeons, and in their behavior they have much in common with the *Columbidae*. For example, sand grouse drink the way pigeons do, thrusting their beaks into the water and swallowing without raising their heads, unlike other species of birds. On the basis of that characteristic, some ornithologists maintain that sand grouse are closely related to the pigeon family.

On the basis of other characteristics, however, and after years of observation, I have come to the conclusion that the sand grouse are related to an entirely different group, the *Scolopacidae*, a family that includes the redshanks and snipe. But enough of technical questions of classification. We are more interested in the biology of these birds, which in many respects is unique in the avian world.

Many species of sand grouse occupy the arid regions of Eurasia and Africa. Since they nest in the remotest and most inaccessible areas on earth, and since, in addition, they are extremely shy and secretive in their behavior, little has been known about them. Then, in the mid-sixties, scientists in the Republic of South Africa—the ornithologists Thomas J. Cade and Gordon L. Maclean—began studying the habits of the species of sand grouse that live in the South African desert. The results of their investigations so intrigued me that I decided to go ahead with a plan of my own that I had been thinking about for several years. Since 1964, when I first encountered the sand grouse on a journey in the Sahara, I had wanted to make a special study of them. If the species in the deserts of South Africa displayed such interesting characteristics, what would their relatives in the even more hostile environment of the Sahara be like?

At the time, hardly anything was known about the life of the Sahara species. In fact, some varieties had rarely been seen by scientists and were known only to the native nomads.

After several weeks of careful preparation, my wife and I set out, early in 1968, on an expedition that was to prove far more strenuous than we had dreamed. And if any outsider had predicted some of the results of our researches, we would have laughed at his wild fantasies. On the other hand, had we known the difficulties of the project in advance, we might well not have undertaken it. The planned one year turned into three, during each of which we spent nearly five months at a stretch in the Sahara.

The operation was a gamble from the start, for there existed virtually no preliminary data on the Sahara sand grouse, nothing on which we could build. Since we had no inkling, when we began our work, of the very special demands that some species of sand grouse make on their biosphere, we could theoretically have chosen any region within the huge desert, any part of an area of more than 3.5 million square miles.

We knew from the start, however, that our investigations would take several months. That made us dependent on a fixed base in an oasis, where food and water would be available. We chose a tiny military fort in the northwestern Sahara for our base. Along with the indispensable water and protection from the weather that the fort offered us, it also seemed a good bet because on one of my earlier visits to the Sahara I had once observed sand grouse not very far away.

The fort is a romantic-looking desert compound built by the French at the end of the last century. When that part of North Africa gained its independence, the French marched out and Arab soldiers marched into the fort. Anyone who had not been aware of its existence would not have recognized it as a fort until directly in front of it. Almost the entire length of the dried mud wall that surrounds the compound is covered by yellow sand dunes that the wind has piled up since the fort was built. Because of the dunes, which would allow an attacker to climb right over the walls, and because of the nature of modern arms, the military importance of the mud fort is just about nil. The chief occupation of the forty soldiers who, with their families, lived there during our stay consisted in the ritual of raising the flag on the small flagpole every morning and lowering it every evening. Besides that demanding activity, the soldiers carted away wheelbarrow load after wheelbarrow load of sand from the dune that kept forming in front of the fort's single entrance and dumped it in the desert. Since the wind often carried the sand back to the entrance faster than it could be removed, the task was by its very nature frustrating. All the more amazing, then, that in the three years we lived in the fort this duty was never neglected.

The fort was also the residence of the caid, who was responsible for the civil administration of a desert region of more than 7,500 square miles—but with a population of no more than a thousand. The helpfulness and hospitality of the caid contributed enormously to the ultimate success of our undertaking.

Aside from the soldiers' rifles, the wheelbarrow, and a radio that maintained contact with the outside world, the most important token of civilization was a diesel motor that supplied current for the four bare electric light bulbs that "illuminated" the fort. One hung at the entrance, the second in the caid's living room, the third in the room that was kindly placed at our disposal, and the fourth from a wire in the center of the fort, near the flagpole. That one no longer burned, however, for the socket had filled with sand. The diesel motor was started at sundown and turned off at 10:00 P.M., to our great relief, since it had no muffler. On the other hand, the frightful noise it made had a practical use. If we were overtaken by darkness on one of our excursions in the desert, we could hear the noise of the motor as much as 6 miles away, and it served to guide us back.

I shall never forget the first night of our stay at the fort, when the motor died down at ten o'clock and the light of the bulb slowly went out. In our room, the glorious stillness of the desert night was replaced, within minutes, by a curious noise that I had never heard before. It began very softly in one corner of the room, a sort of crackling and scratching. Soon the noise became a thousand times louder, until it filled the entire room. When I lit the gasoline lantern we saw a horrible sight. The stone flags of the floor could hardly be seen. Thousands of huge black cockroaches covered it. The insects were swarming out of every crack in the walls and floor. They probably fed on the large quantities of grain that were stored in some of the rooms of our "apartment." For the moment we were helpless to do anything against such numbers. Fortunately, the cockroaches could not climb the smooth metal legs of our beds. By the first gleam of light the next morning the horror was gone. And that night we tried to get a grip on the situation.

First we attempted biological pest control by keeping a desert fox, or fennec, and two desert hedgehogs in our room. Their appetites were so great that they ate fire lanes in the advancing hordes of cockroaches. It sounded as if someone were trying to gobble down a whole bag of potato chips as fast as possible.

When, in spite of our unusual pets, the insects seemed to increase rather than diminish, we resorted to chemical pesticides. The morning after we set out our poisons, we were able to shovel 90 per cent of the cockroach population into a wheelbarrow and take it out into the desert. Once the number of insects was reduced to reasonable proportions, our pets managed to keep them under control.

Our somewhat precipitate decision to come to the Sahara appeared to have been a fortunate one, for this year the sky was cloudier than usual and the possibility of rain seemed strong. Clouds steadily moved down from the north, from the Atlas Mountains, in the form of a cold air front advancing into the desert. Each forward movement of clouds was preceded by a heavy storm of sand and dust that kept us prisoners in the fort for days on end. The storms blew day and night, and the sun was completely obscured. Frequently it got so dark inside that we had to light the gasoline lanterns. The air in our room filled with fine dust so dense that the rays shining through the lanterns' facets and crudely polished glass formed geometric patterns in the air. Sand and dust penetrated the room through cracks in the windows and doors. The small sand heaps that formed on the window sills grew until the sand trickled down onto the floor and there, too, formed rapidly growing heaps.

On days that were free of sandstorms, we spent many hours looking up at the sky and watching the rain that fell in broad, dark veils from the clouds. Each time we hoped that some of the rain would finally reach the parched surface of the desert, and each time we were disappointed. It always evaporated several feet above the ground, as soon as it encountered the layer of hot air. The cloud ceiling was simply not thick enough yet to screen out the radiation from the sun, so intense that a layer of heat continued to form along the ground.

But one day it happened. Up to early afternoon several "ghost rains" had fallen. The coolness resulting from the evaporation of the rain gradually dissipated the warm air layer, and for the first time since we had arrived two weeks earlier a heavy shower of rain reached the ground. The spell was broken. It rained all night and into the afternoon of the next day. Then, within an hour, the sky cleared again, and so it remained, probably until the end of the year. According to the natives, it had been the first rainfall in five years.

Here was a rare chance for us to observe an entire life cycle based on rain water. The ground in the hollows of the dried riverbeds was soaked to a depth of a foot and a half, and that certainly was enough to start plant growth.

We set to work at once and mapped out a study area some 6 miles wide and between 35 and 40 miles long, running north and south. The rain had fallen over all 225 square miles of the area, within whose boundaries were a great many different types of desert. Before three days were out, a green gleam could be detected in the wadis. After ten days many of the plants were already about 8 inches high and in full bloom. A considerable number were spurge (*Euphorbia gyonina*). The ribbons of vegetation in the wadis ran through the desert like a system of green veins, and one day sand grouse appeared. They came by the hundreds, their penetrating flight call of *quito-quito* seeming to fill the air. These were spotted sand grouse (*Pterocles senegallus*). First they flew over the area of plant growth at a great height and at great speed. Our binoculars gave us a good look at the typical appearance of these birds in flight. They have long pointed wings and a long forked tail formed by greatly elongated, narrow tail feathers. The tail serves the bird as a rudder when it is in swift flight. We could also see the black stripe bisecting the light-colored belly, which is the main distinguishing mark of the species. By our calculations, the birds flew at a speed of over 60 miles an hour. The high speed in itself represents an adaptation to desert life, since the birds must cover vast distances to find patches of vegetation.

The birds flew over the green-tinted wadis several times. Each time they passed lower over the new plant growth, and finally, with loud cries, they dropped down into a huge stony area. The moment they settled to the ground, they became almost invisible to the human eye. The plumage of the sand grouse is perfectly matched to the ground colors of desert rocks. Only when the birds began to move again, toward the nearest wadi, could they be recognized. Walking on their short legs over the rocky desert floor, the sand grouse looked like nothing so much as moving stones. For almost the entire day the birds stayed in the wadis feeding exclusively on the fresh seeds of the spurge, which had already formed in the short time since the rain. Evidently spurge seeds are a favorite food of the sand grouse.

Although the rain had fallen equally over our whole study area, the spurge plants grew abundantly only in the southern part of it. More and more flocks of sand grouse discovered this island of vegetation, and within a short time several flocks, amounting to about 400 birds, were swarming over the wadis. Soon the chief foe of the sand grouse put in its appearance: a pair of lanner falcons. The better to prey on these extremely nervous birds, which constantly peer upward even while searching for food, the falcons had adopted an ingenious method of hunting. They flew with very shallow wing strokes so close to the ground that they almost touched it, and sand grouse in the lower-lying narrow wadis were unable to catch sight of them in time. In addition, the low-flying falcons, because of the gray and brown coloration of their backs, harmonized with the color of the desert.

As a protective measure against that kind of attack, the sand grouse have developed a highly differentiated mode of social behavior. One of the flock constantly circles over the area at a great height keeping watch for approaching falcons. A remarkable aspect of this "observation flight" is the exceptionally slow speed that the normally swift bird contrives to sustain by beating its wings rapidly but without a follow-through movement. It is much like the flight of the meadowlark in song, except that the lark flies straight ahead. The birds hunting for food on the ground are in constant voice contact with their aerial sentinel. If the sentinel bird spots a falcon flying close to the ground, it immediately reports the danger to its fellows by a special call. The sand grouse then huddle down among the stones and remain motionless, trusting to their protective coloration. Several times I was able to observe a falcon flying over a crouching sand grouse without detecting it. When the all-clear comes—again a special cry from the sentinel—the sand grouse resume their feeding. I assume that the birds take turns performing guard duty. Since the sentinel bird may, of course, be attacked by a falcon at any time, this "social service" represents a high degree of risk. The sentinel is probably safeguarded from attack from below by keeping the falcon constantly in sight, and from above by flying at a height of about 650 feet. The bird need only see the enemy in time. If it does, it can escape. According to my observations, the lanner falcon cannot outrace the sand grouse, at least not in straight flight.

It took us many weeks to observe and, more important, decode the remarkable social conduct of the sand grouse. Since sources of food in the desert are widely scattered, the birds display a distinctly competitive attitude toward one another. During their common hunt for food they all keep a definite distance apart. Each bird claims its own feeding ground, a circle about three feet across. If a bird ventures into its neighbor's territory, the latter calls attention to the infringement by erecting its tail feathers in a fan. In most cases, that warning suffices to restore the proper distance between the birds. But if the warning is ignored, the first-comer attacks the intruder with outthrust head and a snarling cry.

Shortly before sundown, the sand grouse fly up from the wadis and a few miles out into the rocky waste. The various flocks join in a great swarm, and together they go seeking places to spend the night. They choose an especially inhospitable spot without a trace of vegetation in the middle of a wide area of loose stone. Here they make sleeping hollows by shifting back and forth and from side to side, as they do when taking sand baths.

Since small mammals like the desert jerboa cannot subsist in these stony wastes completely lacking in vegetation, their enemies, jackals and foxes, do not roam here at night. The sleeping birds are therefore safe in their hollows. They will use the same camping ground for several weeks, though each bird makes itself a new sleeping hollow every evening.

The carnivorous falcons and the many small insectivorous birds of the desert ingest liquids in their food, but the seed-eating birds must have water every day. For sand grouse to exist at all in the desert, they need not only plant seeds but also a reliable water hole not too far away from their feeding ground, a rare combination in the desert.

In the Sahara there are three main types of water hole available to the sand grouse. The first is a hollow in the rocks that temporarily fills with water after a rain. Theoretically, every one of the innumerable holes that erosion has produced in the rocks of the Sahara could serve as a drinking trough. But since rain comes only once every few years in most of the desert, and since the rate of evaporation is very high, this type of water hole is both rare and short-lived. Rain water remains for any length of time in only a few very deep hollows in rocks that lie in the shade of overhanging cliffs.

The second type of water hole is one fed by rains that fall in regions remote from the desert. Many dried riverbeds run several hundred miles into the desert from the rainy Atlas Mountains, which border the Sahara on the north. These riverbeds, often filled with sand, are relics of an earlier epoch in the earth's history, when the climate was moister than it is now. Today the rivers dry up only a few miles south of the mountains in which they arise. But their water continues to flow out of sight in the substrata beneath the riverbed for hundreds of miles. In the middle of the desert, at the lowest points in the bed of the dried river, this ground water comes to the surface again in the form of water holes or small ponds. Many of the animals of the Sahara supply their needs from such "river water."

Finally, some water holes from which the sand grouse drink hold rain that fell from the sky millions of years ago, when the continents were much wetter. This fossil rain water has been stored in low-lying, permeable strata of rock until the present day. Where the strata of rock have been exposed by weathering, the water emerges in the form of small ponds around a spring.

Our sand grouse depended on a water hole of the second type. Bounding our study area on the west was a one-time river valley more than 35 miles long in which we had discovered several water holes. The water came all the way from the Atlas Mountains, far to the north. Every morning at seven o'clock, with such regularity that we could have set our watches by them, the sand grouse flew to a big water hole in a wadi about 2½ miles from their feeding ground. They had a curious way of rallying each other to set out on this drinking expedition with a steady series of cries. Then one bird would take off. Calling loudly, it circled over the flock. The birds still on the ground responded with a faster and faster series of calls, until finally one bird after another, one flock after another, rose into the air, and the whole great swarm, constantly calling, would fly to the water hole.

The sand grouse also maintain an effective sentinel system when they go to drink. In the desert, water holes are crossroads for every form of life. Jackals lurk near them, and nomads regularly come to water their herds. Only in the rarest cases, then, do the sand grouse succeed in flying directly to the water. If the birds detect any disturbance, they turn away at once and drop to the ground perhaps a mile off. Since they probably cannot see the water hole from there— the holes are always found in the lowest dips of the riverbed—a patrolling sentinel soars high and checks the situation. When the danger has passed, the sentinel returns to the waiting flock to announce that the coast is clear, and they all fly off to the water. The drinking itself is done in great haste and frequently takes no more than fifteen seconds. After the birds have drunk, they return by the shortest route to their feeding ground. Unlike other animals, sand grouse seem able to drink water with a high salt content. During the hot summer months, when many water holes dry up, and during periods of severe sandstorms, which fill the water holes with sand, they have actually been observed trying to drink sea water along the Atlantic coast.

We were particularly eager to study the drinking habits of the sand grouse. For this, it would be necessary to camouflage ourselves, since the birds at the water hole were extremely shy and wary, and not the slightest movement escaped their notice or that of the sentinel bird. But there was no high vegetation in which we could hide, and a camouflage tent was out of the question—it would have excited too much curiosity from the nomads.

I hit on the idea of digging a sort of foxhole in the steep embankment on one side of the watering place, a hole deep enough to sit in. That meant several days of strenuous work with pick and shovel. The muddy ground was so hard that I felt as if I were driving the pick into baked rubber. Only a few small fragments at a time could be broken loose. Since I could not dig except during the cool morning and evening hours, I did not finish the hole, which was about 5 feet deep, until the third day. I had chosen a small protruding ridge in the embankment for my hiding place, and I drilled observation holes through the 8-inch-thick outside wall. Through these peepholes I could see the water hole and part of the embankment to my left and right, as well as the bottom of the gorge. A sheaf of dried plants, gathered with great effort in the gorge, helped camouflage the entrance (see Plate 24).

The very first morning after my hiding place was completed, I was

24. The author in the hideout he dug for observing sand grouse at a water hole in the northwestern Sahara. Wet towels are wrapped around his head and neck as protection from the intense heat, which can reach 140° F. in the shade.

seated in my hole, full of tense expectation, long before sunrise. I had quickly bored an additional peephole for the telephoto lens of my camera. In this confined observation post, my legs had fallen asleep by the time I heard the familiar calls of the sand grouse, shortly after seven o'clock. Soon the flock, with great fluttering of wings and constant calling, dropped down close to the water hole. I could not see the birds, but I was able to follow their landing with my ears. Hand on the camera's shutter release, I waited for the flock to come along from the left and appear at the edge of the water hole. They did not come, but, instead, flew up and away in alarm. What had happened? Had they caught a glimpse of the camera lens? Then I saw the "disturbance" appear from the right—in the form of a donkey. It drank and after that, to my consternation, it went to sleep. What should I do? The simplest thing would have been to leave my hiding place and chase the donkey away. But the sentinel bird was already circling over the water hole and the sleeping animal, uttering its unmistakable warning cry. If I crept out and it saw me, I would have betrayed my hiding place for good to the wary birds. So I tried something else. I used the long, wide rubber band that ordinarily

went around the lenses in my camera kit to make a slingshot, and through one of the observation holes I bombarded the donkey with small pebbles. No one need think I hurt the beast. Twenty direct hits, and it did not even awaken. I stayed fruitlessly in my hiding place until ten-thirty. After four and a half hours—and when the sentinel bird had not reappeared in more than an hour—I concluded that the birds had gone to another water hole. The donkey was still asleep in the same spot. I had to keep a firm grip on myself to refrain from waking it from its dreams in the ungentle manner the nomads use.

Next morning, with fresh hopes, I was again installed in my hiding place. The sun had already risen, and I was sure the sand grouse would appear at any moment. Then, through my left peephole, I made a crushing discovery. A nomad woman singing a shepherd song was walking along the edge of the embankment, hunting in the riverbed for vegetation that she could use as fuel in her small mud stove. Steadily she approached my hiding place. I heard the sand grouse, and again I could hear them turning away. I still hoped the nomad woman would not notice the hiding place and would walk right by. No such luck. Even while I was trying to think what to do, she discovered the remarkable collection of firewood at the entrance to the hole. Looking up through the stalks, I saw the woman's joyful face and her outstretched hands. I forestalled her by parting the roof of stalks and roots and sticking my head through, at which the joyful expression gave way to terror. She uttered a shrill scream and ran away. Then I made a mistake. I jumped out of my hiding place and ran after her to reassure her. When the unfortunate woman, who possibly had never seen a European before, saw that she was actually being pursued, the distance between us increased perceptibly. She disappeared behind a chain of hills more than half a mile away. My reddish blond hair and my beard, along with a newly acquired sunburn, must have added to her horror.

As I jumped out of my hiding place, I saw out of the corner of my eye that the sentinel bird was watching the whole episode and that it reacted with at least as much alarm as the woman did. From that day on, the hole was an infernal place, to be avoided by both nomads and sand grouse.

It took us two days to discover the alternative water hole the sand grouse were now using, and another three days to prepare a new observation post. At last, fourteen days after I had begun to dig my first hideout, I could watch the spotted sand grouse drinking and was able to take my first photos to document the process.

The sand grouse had come to our study area at the beginning of March, after the rain. It was now the middle of April, and already there were days when the temperature rose to more than 35° C. (95° F.) in the shade. Day after day, week after week, for as long as ten hours a day, we scoured the country trying to find a sand grouse nest. For hours we sat on low hills and with strong binoculars or a telescope on a tripod searched the stony surfaces that stretched to the horizon, looking for solitary birds that might have separated from the flock in order to brood. We discovered nothing. Our morale sank to zero. The flocks continued to stay together, and, as far as we could see, they had not diminished. How could this be? Why did the sand grouse not take advantage of the spring season, with its comparatively lush food supply, to brood, as other birds of the desert did? Many species had already finished their brooding period and were busy raising their young.

The vegetation was already beginning to parch, and from one day to the next the green ribbons of plant growth in the wadis turned yellower. The spurge had cast its last seeds and was shriveling in the growing heat, its stalks contracting toward the center. The spreading, bushlike plant was gradually assuming the shape of a bottle. The outside stalks shielded the green stalks in the center of the plant from the sun's rays and protected it from drying out, enabling the plant to prolong its life by several weeks. Since the wind carried away most of the dry seeds, food for the birds grew steadily sparser, and competitive behavior among the feeding sand grouse flocks constantly increased. Each bird claimed for itself a larger feeding range, and the birds began to fight more frequently. But now the males displayed a striking inhibition about pecking at intruding females. Probably this behavior gradually leads to the formation of pairs. In a few days the flocks of birds finally broke up, and more and more single pairs were seen hunting through the wadis for the last available seeds. Other distinct changes could be noted in the behavior of the birds. They were shier and more secretive than ever; at the slightest disturbance they immediately pressed themselves down among the stones, becoming even more inconspicuous than they had been before. In short, their behavior was typical of birds trying not to betray a nesting place.

We were both delighted and astonished at what we were seeing. Only now, when there was virtually no food left and the fiery hot Sahara summer was about to descend, were the sand grouse beginning their mating season. Then, just as we were deluding ourselves that the goal of our efforts was near, that after a five-week vigil we had at last found a sand grouse nesting place, we made a shattering discovery. We saw that the bird population at the water holes was diminishing day by day. After their morning flight for water, several pairs would not return to the feeding ground. Instead, they flew away northward at a very high altitude. One pair after another vanished among the endless wastes of the Sahara.

The departure of the birds was accompanied by a change in the weather. The start of summer in the Sahara is announced by a change in the direction of the wind. Now, in the middle of April, the northerly winds gradually shifted around to southerly. The high winds that blew from the interior of the Sahara were only warm as yet, but they gave us a foretaste of the fiery breath of the great desert that would take our own breath away six months later.

One day, when no more birds came to the water hole and the stony wasteland was utterly barren and deserted again, we, too, left the southern part of our study area.

The sand grouse's mysterious conduct, different from that of any other species of bird inhabiting the desert, had made us more eager than ever to solve the puzzle of their brooding places. But how were we to go about finding these perfectly camouflaged birds, which, besides, had such secretive habits? Our chances of spotting them in the inconceivable vastness of the desert seemed extremely slim. The only clue we had was our observation that the birds had flown off in a northerly direction. Also, we reasoned that the paired birds would still have to be fairly near a water hole.

Some 30 miles from the fort, in the northern part of our study area, we set up another research camp. It was pointless to search hundreds and thousands of square miles for single pairs of birds, so we decided we would first try to locate water holes where they might come to drink. If we found those, and saw that they were being visited by sand grouse, we could then try to locate the birds' nesting places. But by now, the second half of April, most of the water holes were already dried up. If the sand grouse were nesting, they would need a water hole that could supply them for another two months, deep into the Sahara summer. Brooding their eggs and raising their chicks would take at least that long.

After a three-day search, we actually discovered a large, deep water hole that met those requirements. It was in the bed of the same dry river that fed the first water hole, 30 miles to the south. We were so excited that we dispensed with sleep for that night, and the next morning, half an hour before dawn, we were dug in on a small sand dune 1,000 feet from the water hole. When the sun rose above a distant mountain range to the east, our suspense increased to the point of nervous irritability. Would any birds come to drink, and would we succeed in recovering contact with our vanished flock of grouse?

At eight o'clock three pairs dropped down by the water hole, one by one. And so the second and most exciting phase of our work began: our investigation of the nesting biology of the sand grouse. The instant the birds took flight again, we dug ourselves out of the sand as fast as we could and ran up the steep embankment to watch them flying off. Because of their speed, we had only a moment to see that they were flying back in the direction from which they had come, eastward. Almost immediately they were swallowed up in the glaring white brightness of the morning light.

Next morning, when four pairs reported to the water hole, the problem was the same. It was utterly impossible to follow the flight of the birds with our binoculars. Moreover the terrain cut us off on all sides. Folded slabs of cliffs 350 feet high, many of which had weathered to huge rubble heaps that looked like gigantic coal piles, ran in all directions, blocking the view. In addition, by this time of year the air above the black rocks was so heated that it shimmered violently. The flying birds vanished in those wavering layers of air, which refracted light so strongly that objects were distorted beyond recognition. The effect was as though the birds were plunging into the water.

To find the nesting places of the birds, in spite of all these obstacles, we developed a time-consuming and tedious method. First we drove around an 80-square-mile area to the east of the water hole and made as exact a sketch of the terrain as we could. Armed with this sketch, we began charting the birds' return flight from the water hole. With a compass we determined the exact direction of the flight and entered it as a line on our sketch. Day by day we moved our observation point approximately a third of a mile along the line of flight in the direction of the presumed nesting place. In this way we could connect the individual flight paths of the birds despite the light's flickering through the strata of air and the intervening ranges of low hills. Four days of this section-by-section movement brought us out of the hilly area, and before us there now stretched a flat, stony wasteland extending to the horizon. The surface consisted of brown pebbles as small as peas and as large as walnuts. Interspersed with them were small dark stones, which gave the whole vast expanse a speckled look. In this open country we could keep the birds in sight longer, and so we doubled the distance we moved our observation post each day. We were interested to note that the line of flight growing longer and longer on our sketch ran with amazing straightness. This permanent flight route seemed to be proof that the birds always chose the shortest distance between two points. One point, which we already knew, was the water hole; the unknown point must be their nesting place.

Our suspense reached its climax when, on the morning of the sixth day of our tracking operation, we saw a pair of sand grouse only about 300 feet from our observation point just starting on their morning flight for water.

In half an hour we heard the cries of the returning pair. The birds dropped down in almost exactly the same spot from which they had flown up. We could see with our binoculars that immediately after landing they ran toward some goal not yet apparent to us. The female, whom we could distinguish from the male by her special coloration, settled down on a spot marked by a chance accumulation of several sizable stones. Such groups of stones were rare and tended to be scattered miles apart in the stony waste. They were the last remnants of strata of hard rock that had resisted the forces of weathering.

The moment the female crouched down in the midst of the stones, which were about the same size as the bird, it became almost invisible so well was it camouflaged by its protective coloration. The male continued on. While my wife kept focusing on the collection of stones, I walked slowly toward the female. The astonishing thing was that it did not fly up, although I had considerably infringed on the "escape boundary" of these normally supershy birds. Instead, the male flew into the air, toward me, and landed no more than 100 feet away. It then performed an act most peculiar for a desert bird that counts on camouflage for concealment. The male raised its wings vertically to show their white underside—it was like a waving flag seen from far off. Then it took little leaps, all the while violently beating its wings. It was clear that the male was trying, by this display, to lure me away from the female, to which I was getting closer and closer. But I did not fall for the diversionary maneuver and was able to approach within 30 feet of the female. Never before had I been so close to a sand grouse without myself being hidden. The bird merged perfectly into its surroundings. If I turned to look toward my wife for only a moment, I had to peer hard in order to find the female again. The coloration of its plumage, brown with dark spots, exactly corresponded to the colors of the stones. It sat perfectly immobile on the desert floor, its head withdrawn. Obviously it trusted to its protective coloration. There could be no doubt that the bird was incubating eggs. In order not to frighten it, I went no closer, but walked around it at a distance of 30 feet, pretending not to see it and taking care not to look directly at it. I knew from experience that many desert birds, when they are brooding eggs on the ground, feel discovered if eye contact is made and will flee.

Next morning, when the sand grouse pair again flew off to the water hole, we seized the opportunity and quickly went over to the collection of stones. On one of the last days of April, after forty-six days of intensive searching, we had reached our goal. Here, almost four miles from the water hole, was the nest of a speckled sand grouse. If it could be called a nest. It was nothing but three eggs lying in a shallow hollow among the pebbles, with no nesting material at all. Even the eggs were perfectly camouflaged by their shape and color. They did not have the characteristic, easily recognizable shape of an egg, with a plump round end and a slender pointed one. These eggs were almost equally plump and rounded at both ends. They rather resembled a short, cylindrical sausage. With their brown and spotted coloration, which exactly corresponded to that of the surrounding stony desert, they looked more like pebbles than eggs (see Plate 25).

Here at last was proof that the sand grouse had not, after all, brooded unnoticed in the area where the food supply had been greatest. The southern part of our study area, with its burst of plant growth, had been only a prebrooding biotope in which the birds gathered and subsequently paired off. The brooding area, 30 miles to the north, was much drier and more hostile to life. Nor was this stony terrain any cooler than the feeding grounds. On the contrary, it was a great deal hotter. So it was hard to fathom why the birds had moved to such an inhospitable region.

We could not tell from the eggs how long the birds had been brooding. But most of the sand grouse had left the feeding grounds in the middle of April, so the eggs could not be more than two weeks old.

On May 1, the long Saharan summer that we so feared began. The temperature climbed day by day, and our work in the study area became steadily harder. For two weeks there was not the slightest indication that the period of incubation was nearing its end. During this period, while we waited eagerly for the next great event, the hatching of the young, we had ample time to study the incubating habits of the birds.

A few days after we discovered the first clutch, we found a second, with two eggs. Next day a third egg had been added. That gave us the starting point of incubation, so that later we were able to calculate its exact duration.

Eventually we discovered a total of twenty sets in the area of desert wasteland we were surveying. Pairs frequently nested at distances of several miles. The shortest distance we found between a set and the water hole was just under 2 miles, the longest 5 miles.

We particularly wanted to know why the birds nested in the midst of a chance accumulation of sizable stones. Our theory was that they used the larger stones as landmarks to help them find their own eggs on the flat and unchanging stone surface. But that explanation satisfied us only until the day we found two sets that were not surrounded by a ring of stones. The nesting hollows holding those eggs lay in a perfectly flat region of small pebbles. Even from the top of the automobile roof we could not see the least sign of a landmark that the birds might use. Nevertheless, every morning when they returned from drinking the birds invariably found their own eggs without effort. Obviously the large stones had some other significance.

We did not solve the enigma of the stones until after two full weeks of close observation. In the meantime, we studied the hunting methods of the lanner falcons, which had also arrived at the nesting area, and noted still another enemy, a pair of brown-necked or desert ravens, hanging around the brooding area. The ravens were after the eggs or, later, the sand grouse chicks.

Both the falcons and the ravens had brooded in March and April, and had raised their young while the supply of food was ample. The falcons feed their young chiefly on migratory birds, which in the spring cross the Sahara in great numbers en route to Europe from central Africa. Worn out from the Sahara crossing, many of the birds fall an easy prey to the falcons.

In summer, when the food supply dwindles, the only prey left to

25. A spotted sand grouse's clutch of three eggs next to two large cover stones in the heart of the rocky desert

the falcons, the ravens, and their newly independent fledglings is reptiles and Saharan birds, especially the sand grouse and their young. Small mammals are rarely available because they lead a largely nocturnal life. The selection pressure exerted by the falcons on the sand grouse is most intense.

As far as we could observe, the falcons made no attempt to attack the sand grouse during their morning flight to water. That astonished us at first; for although we had seen that the falcons could not catch the grouse in straight flight, several times at the feeding grounds we had observed them plunging from a great height at the flocks of sand grouse, in a sort of power dive. In those dives they attained a flying speed of up to 100 miles an hour, which gave them some chance to catch a sand grouse. But that method of hunting seemed to work only when the flocks were very large; it was probably of no use against individual pairs flying to and from the water hole. In the brooding area, the falcons had developed a new technique aimed at surprising and striking the sand grouse on the ground.

During May and June both falcons and ravens can be seen on the desert floor in the heart of the brooding area. They watch the pairs fly off to the water hole, but do not follow them (see Plate 19). The falcon waits and flies up to attack only when the sand grouse pair have returned from drinking and landed near their eggs. The falcon uses the same technique we had observed earlier at the feeding grounds—flying low, close to the ground, in order not to be seen by the sand grouse.

A sand grouse would immediately detect a high-flying falcon and would freeze into immobility, so that the falcon would be unable to see it from above. But a falcon flying close to the ground stands a good chance of catching a sand grouse by surprise. For in their brooding areas the sand grouse do not use their highly effective system of aerial surveillance by sentinel birds.

I appreciated the worth of that method of hunting only when I lay flat on the ground, with my eyes at about the same level as a low-swooping falcon's. From a height of about 8 to 12 inches the sand grouse, distinctly silhouetted, stood out clearly on their level brooding area. Often the birds are the sole elevation on the pebbly surface and can be seen by a falcon even when, aware of the enemy's pres-

26. A broody spotted sand grouse female, for once without the shelter of cover stones

ence, they resort to their trick of crouching motionless on the ground.

The attentive reader will wonder how the falcons can make out anything at all in the heated layers of air above the ground, which must be quivering wildly, blurring all contours, even at this early morning hour. The reason is that the blurring does not occur in this case. If the sand grouse's silhouette is the only elevation against an otherwise flat horizon line, it cannot be distorted by other elevations. On the contrary, the layer of hot air serves as a kind of magnifying glass; the distortion makes objects appear larger than they really are.

However, since all the life forms on earth live in a continual reciprocal relationship to one another, and since each kind of behavior on the part of a hunter results in responsive behavior by the prey, the sand grouse have developed a means of protecting themselves against the special hunting method of the falcons. They use those conspicuous accumulations of stones in the midst of which they brood as a sort of visual shield against the low-flying falcons and desert ravens. Within the shelter of these "cover stones," the brooding bird no longer shows as a silhouette against the horizon. When it tucks in its head it cannot be distinguished from the rounded stones, many of them about the same size as the bird itself (see Plate 27). Also, the vibrating hot air has the expected effect of making the outlines of the bird's body appear to merge with the stones. The fact that some sand grouse brooded without this visual shield might indicate that the use of the cover stones is a fairly new development and also provides a clue as to how the new brooding behavior might have arisen.

The brooding bird is probably unconscious of the advantageous optical effect. Its behavior has arisen by natural selection, in the sense that those of the species that brood among stones have a better chance of surviving than those that do not. The choice of such a nesting place certainly improves their chances of reproducing. It can be predicted that ultimately the falcons and ravens will learn to see through the trick, and then the sand grouse will have to invent some other protective mechanism.

When the birds returned from their drinking flight in the morning, the female always ran full tilt to its eggs to protect them from the rapidly increasing temperature and from the sun's rays. In the desert the female's task has little in common with real incubation. In cooler climates, brooding birds use their bodies to keep their eggs at an incubation temperature that is generally higher than the temperature around them. In the brooding area of the sand grouse, by the middle of May the air just above the ground reaches 40° C. (104° F.), and the temperature of the rocks can rise as high as 55° C. (131° F.). The eggs cannot endure a temperature much higher than 40° C. The female must therefore be able to cool rather than warm them.

But how can a warm-blooded animal keep a clutch of eggs "cool"? From the day we first raised that question, more than a year passed before we were able to solve the enigma of "sand grouse refrigeration." I shall deal with it later.

Early in the evening the mates change places. The female leaves the eggs while there is still an hour's daylight to go hunting for food, and the male takes over the clutch. It sits on the eggs until early morning, just before the pair fly together to the water hole. Only the male seems to be able to keep the eggs warm during the cool desert nights—a highly interesting division of labor resulting from the extreme temperature difference between day and night.

Every morning I checked the eggs while the pair were away at the water hole. One morning during the third week in May I heard a soft peeping from one of the three eggs. Two days later, I deduced from the behavior of the parents that the first chick had hatched from the egg. The male flew off alone for water while the female remained sitting on the eggs, probably to protect this first chick and give it a sand bath. When the male returned, it hurried to the nest and took the female's place; the female then flew off to the water hole alone.

If this did indeed mean that the first chick had hatched, the others would be hatching at the same interval at which the eggs had been laid. We guessed that the interval was probably one day. Our theory was confirmed on the evening of the third day. At seven o'clock, half an hour before sundown, the female left its brooding place. Through our binoculars we spotted three tiny chicks following it in formation. The male joined them at once, and the family was complete. We watched with the keenest excitement. For weeks we had been asking ourselves how the chicks would be brought to the water hole to drink. It was unlikely that the parents could solve the drinking problem by carrying water in their crops. A sand grouse does have a capacious crop, in which it can store a large supply of water. But the amount just suffices for the considerable daily needs of the adult bird in its hot environment. The bird probably pants away most of the crop water to cool its body. Of course, it was conceivable that the adults would fly to the water hole several times, first to fetch water for the chicks and then to cover their own needs. But the construction of the crop does not permit these birds to regurgitate water. More-over, the distances flown in the hot desert must be kept to a mini-mum because of the exertion connected with flying.

Besides its water, the dry riverbed seemed to offer the birds other benefits. Most of the sparse vegetation grew at the bottom of it, and so a great deal of nourishment in the form of seeds was available to the sand grouse family. In addition, the bushes would provide shade during the hot part of the day.

But the brooding place for our pair was nearly 4 miles from the water hole. We simply could not imagine how those tiny chicks, with their short legs—which made them even worse walkers than their parents—were going to cover such a distance in the blazing desert. Never mind that a moving sand grouse in terrain without cover was utterly exposed to its enemies. Now, at the end of May, the tempera-ture was already more than 40° C. (104° F.) in the shade. And there was no shade in this open area. The sun's rays, with no clouds to filter them, mercilessly seared the stony waste for fourteen hours a day, heating the rocks to as much as 60° C. (140° F.). If the family some-how intended to cover most of the distance on foot—which, accord-ing to our observations of their walking speed, would take between ten and twelve days—how were the chicks to quench their thirst during the long hike, and how was the family ever to cross the black, fiery hot, and rugged chain of hills that lay between the nesting site and the bed of the dry river, with its water hole? Even the water hole itself lay in a deep declivity and was hard to reach except by flying.

Seeing the sand grouse family choose the evening to leave their nesting site for the first time stirred our imaginations all the more. Various new possibilities occurred to us. Would the family try to reach the water hole by night marches, avoiding the heat of the day as well as the falcons and ravens? It was not so strange an idea. Might it not be that the chicks could quench their thirst during the night walk by sucking up the tiny amount of dew that is present at night in the desert?

We quickly noted that the parent sand grouse behaved with ex-treme shyness and distrust while they were taking care of their chicks, and we were not able to come any closer to them than 200 feet. If they really were setting out on a night march, it would be almost impossible for us to keep them in sight, even in the moonlight, because of their perfect protective coloration. And once we lost them in that vast stony waste, the chances of finding them the next morn-ing were slim. In addition, we had now spent three and a half months in the hot, dry climate of the desert, and our own energies were running down. We were often tramping about from five in the morning to eight at night with only brief interludes of rest.

The sun had set and dusk was spreading over the desert. The huge dark bow of the earth's shadow rose inexorably higher in the eastern sky. It was as though a vast bowl, the black inner surface of a hollow hemisphere, were gradually closing over the globe. The sand grouse family had covered only about 165 feet since it set out an hour ago. We could still make out the birds with our high-performance glasses, but just barely, when we saw that the family seemed to be pausing for a rest. Then something happened for which we had made no place in our speculations. The female began scooping out a sleeping hollow. The chicks disappeared into the hollow, and the female settled down over them to keep them warm during the cool night. The male moved into its own hollow nearby. When the idea that the bird family might undertake a night hike had occurred to us and we had decided to try to follow them, we had fetched our sleeping bags from our camp, which was not far away, and had refilled our water bottles. Though the nocturnal hike was obviously not going to take place, that precautionary measure proved to have been a sensible move. In order to keep our eyes on the sleeping sand grouse family, so that we would not lose them in the morning, we decided to spend the night precisely where we had last observed the family in the gathering darkness.

Within an hour it was pitch-black. It was several hours till moonrise. We lay in our sleeping bags and looked into a black sky dotted with the gleaming points of countless stars. Since moisture is low in the desert, there is almost no refraction and the stars do not twinkle. These hours of the desert night, unsoftened by mild moonlight, have always seemed to me somehow terribly lonely, cold, and cruel. The immobility of the stars in a desert sky conveys a sense of the vastness of the universe. That night I did not feel as though I were looking up into the universe, but as though a new dimension had become visible and lay beneath me. I had the dizzying sense of plunging downward into infinite space. The merciless sensation of the coldness of space, of the vast void out there, was intensified by the absolute stillness of the desert night until it became inhuman, unbearable.

I think I would have screamed had I not at that moment remembered the sand grouse family sleeping so near. Immediately the fate of those birds assumed tremendous importance to me. Now it was more than scientific interest that linked me to them. Rather, my eagerness to learn how and where the birds would obtain their water sprang from a kind of deep inner connection, a solidarity with their utterly different form of life. We, the living, were linked in the face of what I felt to be the cruel coldness of the universe and the emptiness of the desert around us.

Those feelings were perhaps a reenactment of the primal fear that, millions of years ago, had entered the consciousness of some living organism, maybe in the heart of a desert. Perhaps in the remote past of this earth the same sense of immense loneliness had contributed to the development of social contacts, from which, ages later, the idea of humanism would arise.

Never before had I felt the oneness of all life, of all forms of life, so strongly as I did on that desert night. In man's thinking, ordinarily so confined to his own species, living things have always been carefully sorted into good and evil, useful and harmful, and so on. On that night in the desert I realized that only a single idea of life exists and that all forms of life are merely variant aspects of that one idea, like different interpretations of a single theme. A visitor from outside this earth, not bound by human, species-limited thinking, would have the feeling as he observed our planet that he was looking upon a single, coherent organism embracing the entire earth.

When the first gleam of dawn appeared in the eastern sky, I awoke. It was that curious, hazy time of passage from dream to reality. These early morning hours in the desert always have a glassy blue coolness about them. As the light increased, the familiar contours of the sand grouse emerged gradually from the monotonous void of the endless stony desert.

An hour after sunrise, the female, which had sat on the chicks all night, was relieved by the male. The female promptly flew off alone in the direction of the water hole. When it returned in about half an hour, we waited in suspense for what would happen next. In this heat and dryness, the chicks had to have something to drink, had somehow to be supplied with water. The first hatched chick had already gone three days without water. But nothing at all happened, except that the female now took the male's place in keeping the chicks under its wings, while the male flew off by itself to drink. It had been gone about half an hour when something strange did happen. The female suddenly began to utter a peculiarly sharp, high-pitched note at short, regular intervals. The notes sounded something like *queet—queet—queet*. The pauses between the cries were about three seconds. The call, which we had never heard before from a sand grouse, somewhat resembled the navigational signal given out by a radio beacon to guide ships and planes; the signal takes on a special timbre when the vessel is on the beam. After about two minutes we heard the same sounds from the air, almost inaudible and apparently far away. As the female continued to repeat her call, the second call rapidly grew louder, telling us that the producer of the sound was steadily approaching, although we could not detect another bird, even by sweeping the sky intently with our binoculars. The female's call grew louder and more excited. Then suddenly we spied the male, which had returned from drinking and had landed, also calling loudly, no more than 100 feet from the female.

Over the next fifteen minutes we were witnesses to one of the most fascinating adaptations that any living organism has developed in the struggle for survival in the rocky part of the desert. The adaptation represents behavior unique among birds.

The male ran toward the female at a speed most unusual for a sand grouse, both birds meanwhile keeping up their special call. We hardly recognized the familiar male. Its breast and breast feathers were so puffed out that the bird looked completely deformed. It seemed greatly overweighted in front and several times almost fell on its face (see Plate 30). About 15 feet from the female, the male stood still. Constantly calling, it changed its normally crouched, horizontal posture, straightening up and spreading its wings to bare a large area of dark plumage on its breast. Through our powerful telescope, mounted on a stand, we saw that this area of dark feathers glittered in the morning light. The male had brought water for the chicks in its abdominal feathers, and the glitter came from the thousands upon thousands of tiny water drops.

The erect posture and the sight of the wet, dark plumage of chest and belly immediately lured the three chicks out from under the female. They ran as quickly as they could toward the "water source" being offered them. As they ran, they uttered—very softly, but quite audibly—the same special cries that the male was continuing to voice. These stopped only when the three chicks reached their father. They ran under the bird's belly, stretched their necks up, and plunged their beaks deep into a long, clearly visible channel that ran down the center of the water-soaked feathers from front to back. To keep from being knocked over by the pressure of the three thirsty chicks, the male propped its tail feathers against the ground. Evidently the chicks were sucking water from the male's wet belly feathers. The male's upright posture—we later came to call it "the watering posture"—was probably what made the water flow into that special channel among the feathers, which acted like a rain gutter. Also, the male's cramped posture indicated that it was contracting the muscles of its breast and abdomen to force the water into the channel (see Plate 31).

Now we realized the meaning of those special constant calls that the female and later the chicks had exchanged with the male as it returned from the water hole. They were contact calls, location calls. The female probably began them when it heard the cries of the male a few miles off, before they were audible to us. The calling between the birds might be compared to the signals from a radio beam establishing contact between an airport and an incoming plane, which enable the plane to land even when visibility is poor.

The loud and high-pitched cries of the sand grouse can be heard over a great distance. They allow the male to find its way back to the chicks by the shortest route and in the shortest time, without any protracted searching or flying here and there. That is essential to keep the precious freight of drinking water from evaporating in the hot and extremely dry desert air.

The transformation of the feathers, which normally are water-resistant, into a water-storing device represents a truly astonishing adaptation. Only a few days later we were able to find out how the water-bearing plumage is constructed, and how the birds take in, store, and transport water, when we briefly captured a male for the purpose of examining its feathers. We supplemented what we learned then by watching the male sand grouse's behavior at the water hole during the time the chicks were being reared.

One of the characteristics of a sand grouse is its unusually thick plumage. Under the top layer of breast and belly feathers is another very thick layer of feathers that grow in a peculiar manner. Each one of the fine hairs growing close together is twisted around another to form a pair, leaving the feathers full of minute air spaces that serve as good insulation from either heat or cold. This construction provides a high degree of firmness as well.

Some species of sand grouse also have unusually thick skin over the belly. Unlike the skin of other species of birds, it is not directly connected with the muscles, but is separated from them by air chambers. The dense plumage, the thick skin, and the air chambers—characteristics already present in the newly hatched chicks—form a highly effective insulating layer, which enables the sand grouse to walk about on the stony surface of the desert even at a summer temperature of more than 70°C. (158° F.).

The insulation on the bird's underside has little effect on the eggs, since the high ground temperature around the brooding site will be transmitted to the hollow in which the clutch lies. But I have already mentioned that the sand grouse have discovered a fantastic means of protecting their eggs from heat, a method we found out about only much later, in the course of further researches.

In order to observe the male in the act of storing water, we dug another foxhole in the steep wall to one side of the water hole. During the fiery June weather that was work that could be done only in the cool of night. From the hideout I had an excellent view of the male absorbing drinking water for its chicks. First the bird ran to a shallow spot at the edge of the water hole and quenched its own thirst. Then it spread the feathers of its breast and belly so that the underfeathers and the thick layer of down were exposed. It ran partway into the water, as though intending to take a bath. One moistening caused the twisted downy hairs to unwind into a thick, feltlike mass. The down filled with water; between 20 and 30 cubic centimeters can be soaked up and stored by the feathers.

If a specific volume of feathers is compared with, say, the same volume of a rubber sponge, we find that the sand grouse's down has twice the absorbency of the sponge!

To saturate the plumage more thoroughly, the male moved back and forth sidewise, as it does when having a sand bath. Then it took off on the return flight. Surplus water dripped from the belly plumage. As soon as the bird was in the air, it closed the thick, dry outer plumage so that most of the water-storing down was covered, reducing evaporation during the return flight. The distance the male had to cover with this freight of water amounted to about 4 miles. Just as I had thought, immediately after takeoff the male began to utter those peculiar navigational cries.

But back to our sand grouse family and to our observations of the chicks drinking. The operation took about ten minutes. After the chicks had quenched their thirst and come out from under their father's belly, the adult bird took a lengthy sand bath, probably to dry its feathers. Then the family walked to one of the numerous shallow wadis that crossed the rocky waste. The chicks tried to avoid the slanting but already hot rays of the morning sun by walking in their parents' shadows.

By June the sparse vegetation in the wadis is largely dried up and dead; it was hard to believe that the sand grouse would find anything to feed on during the summer months. They pecked chiefly at the dried seeds of annual plants. The interesting thing about the annuals is that, in contrast to perennials, they cannot endure the extreme living conditions of the desert for very long. They do not survive protracted dry spells. In moist years they spring up very quickly from seeds, but as soon as the plants have formed new seeds they die. The seeds, however, which are produced in unusually large quantities, remain viable for decades, preserving the species. The abundance of the seeds enables sand grouse to live through droughts of several years during which no new seeds are produced.

In the wadis the adults stimulated the chicks to eat by constantly feeding in front of them. The parents also dug at the ground with their beaks to uncover old seeds that had been embedded there for years waiting for the next rain. Hardly any spurge, whose seeds are the sand grouse's favorite food, grew in this part of our study area. It seemed likely, however, that great quantities of spurge seeds were carried here from the southern part of the area by the hot south wind that prevails at this time of year. The transport of their food supply had probably been an important factor in prompting the birds to move from the southern to the northern section of the area. But the mystery remained why the sand grouse, unlike any other desert birds, brooded so late in the year, at the beginning of the Saharan summer.

In spite of three years of research, I have not yet found a satisfactory explanation for this peculiarity. I suspect, however, that earlier in the season the seeds of the annuals may be poisonous or otherwise harmful to the tender digestion of the chicks. Many desert plants give off poisons, bitter substances, or noxious smells to discourage animals from eating them. If some such indigestible substance was present in the fresh seeds, the substance might be broken down by the intense solar radiation so that by summer the seeds were edible.

Since large quantities of seeds were available in the wadis, the sand grouse family did not have to cover any great distances in their hunt for food, and during the hottest part of the day, from eleven in the morning until about five, the entire family rested in the shade of small bushes. By evening, when the birds left the wadi—which would be unsafe for them at night—to find a secure place to sleep in the midst of the pebbles, the family had traveled no more than 650 feet.

The parents were extremely vigilant in caring for the chicks. While

one parent guarded and tended them, the other was usually surveying the landscape for possible enemies. Once, from our shallow hideout camouflaged by dried desert plants, we saw a jackal suddenly
appear out of a dip in the ground. The adult birds detected their
enemy at once, and the three chicks, in response to a warning cry
from the male, hid under a small plant. The adults huddled down on
the stones right where they were, trusting to their protective coloration. The jackal trotted past at a distance of only 65 feet without
noticing them. It had continued on about 600 feet from the family
and had vanished from view behind a low mound when something
highly interesting happened. Before the male uttered his all-clear cry,
which would bring the chicks from their hiding place, both parents
took off. Flying very slowly and at an altitude of no more than 20
feet, they followed the jackal for a while. As I watched, I could only
conclude that it was a reconnaissance flight; the birds wanted to
make sure that the cunning animal was not going to circle around
and creep up on them. Only after they had convinced themselves that
it was continuing in the same direction did they return to their chicks
and call them out of hiding.

If an enemy approaches the concealed chicks, either by chance or
because it has spotted them, the adult sand grouse employ another
highly effective tactic to save their young. We ourselves first encountered it when we walked directly toward the family to get a close
look at the chicks. At a warning cry from the male, the chicks hid
under a small bush, and the parents froze where they were, pressing
themselves against the ground. We carefully noted the chicks' hiding
place and continued straight toward it. When we were almost upon
the family, we looked directly at the parent birds, causing them to
feel that they had been discovered. The male took flight with a great
flurry, but came down only about 60 feet farther on, at the brink of
the wadi, where it immediately began beating its wings noisily and
taking little jumps into the air. It was obvious that the bird was
putting on a show in order to lure us away from the chicks. When we
knelt down close to the chicks to get a good look at them—they had
chosen a soft spot under the bush and had half dug themselves in, so
they were well camouflaged—the female ran toward us, approaching
within about 15 feet. It uttered plaintive cries and put on a most
convincing act of being injured. We pretended to be taken in by the
trick and went after the female. But it was not easily caught. We did
not manage to reduce the distance between ourselves and the "injured" female by as much as two feet. Every time we got near enough

to try to seize the bird, it staggered and fluttered clumsily a few feet farther along the ground, and then would slow down for a while, writhing and keeping up its plaint. We pretended to pursue it for more than 150 feet. Suddenly, when the female had lured us far enough from the chicks, it flew up and away in an elegant takeoff. We had to use our binoculars to see that it was flying in a wide arc back toward the chicks.

It is difficult to explain how such a stratagem, which is employed by many species of birds, could have evolved. The experiences of a bird that actually was crippled by a broken wing could hardly have been transmitted to its descendants, since the bird surely would not survive a real injury of this sort, but would be eaten. Perhaps the reader will want to think out this interesting problem someday.

The watering of the chicks by the adult birds' separate flights to the water hole, during which they remained in contact by calling, continued for three days. By then, the voices of the chicks had developed to the point that they could be left alone in a hiding place to take over the vital navigational calls. From that day on, the adult pair flew together to drink and fetch water, although the water carrying remained exclusively the male's task.

Meanwhile, other pairs had hatched out their chicks. One of these pairs, with its chicks, was 5 miles from the water hole. That pair usually flew to drink and fetch water somewhat earlier than the pair we were concentrating on, which was a mile or so closer to the water hole. By the time our pair set off, the other pair—the male laden with water—were already back. They always flew right over the area where our pair's chicks were huddling in their hiding place. Yet "our" chicks never reacted or responded to the signal cries from the other male, although to our human ears the cries of all the water-carrying sand grouse males sounded exactly the same. The chicks replied only to their own father. This must mean that every sand grouse family has its own individual notes, even if we were unable to perceive the fine distinctions.

27. A brooding spotted sand grouse female among large cover stones as seen from the perspective of a peregrine falcon on the ground. Only a shutter speed of one five-hundredth of a second shows the bird and the cover stones sharply. For the falcon, the shimmering of the heated air causes the contours of the sand grouse to merge with those of the cover stones.

28. Sample of the porous, and therefore cooler, rocks on which the sand grouse females brood

In five or six weeks the chicks were fledged and flew off to the water hole along with their parents.

The following year, 1969, our entire study area was a dismal, sun-parched wasteland. Not a trace of a green plant was to be seen in the wadis. Here and there the withered remains of last year's vegetation were still present, but the stalks of the bushes and grasses were completely dried out and no longer swayed in the desert wind. Bleached by the sun, rigid, they were suffering the same fate as the brown and black cracked stone that surrounded them. The sand-laden wind was gradually, inexorably, grinding away the last evidences of the previous year's ample rains. The desert had removed its mask. This year it was showing us an entirely different face. The

29. Male spotted sand grouse with three chicks hunting for food on the barren rock

31 and 31. Male spotted sand grouse with chicks

32. Male of the coroneted sand grouse *(Pterocles coronatus)* with its chicks

33. Searching for a hibernating poor-will in the Mojave Desert

34. Desert annuals in bloom following heavy rains in the Yuma Desert of the southwestern United States and northern Mexico. The speed with which these short-lived desert plants grow and bloom can be judged by the fact that the sand is still pockmarked and damp from the rain that started them growing.

35. Sunlight reflected on the seemingly polished surface of a desert of stones coated with black metallic oxides

burned-out landscape in which we walked searching for "survivors" conveyed a sense of endless abandonment.

The delicate high notes of the male birds' mating song, which had been constantly in the air the year before, no longer sounded. All we could hear was the rattle of dried branches as they brushed against one another in the wind. The population of sand grouse was down from 400 specimens to about 80. Most of them had moved on to look for places more friendly to life. We searched for three months without finding the least sign that any sand grouse had undertaken to raise families in our study area this year. We then began a series of reconnoitering expeditions, at increasingly greater distances, in the hope of finding a brooding territory.

At the end of May—104 days since the beginning of our new expedition—we found a small brooding area deep in the Sahara, 435 miles by car from our main camp. In that small area we found a total of four sand grouse clutches. To our surprise, this year, which was both drier and hotter than the previous year, the birds were beginning to brood two weeks later, on the average, than they had the year before. This meant that the chicks would be raised at the very height of summer.

That summer, when we were working far from the shade and coolness of any oasis, was to prove extremely hazardous for us. We fully experienced the sheer physical stress of a Saharan summer, learned to regard it as an antechamber to hell. What we had previously dismissed as a tall story—that in the summer months the desert wind is so hot it will sear any unprotected flesh—became an everyday fact of life. To survive at all, we had to drink between 12 and 15 pints of water a day.

Never before had we witnessed such sunrises. The moment the first ray of the sun appeared above a distant mountain, we had the feeling that a gigantic magnifying glass had been placed between it and us. It was as if we had been shut up for weeks in a dry sauna. Between us we lost 50 pounds. The afternoons were the worst; the temperature would rise as high as 48° C. (118° F.) in the shade. The sense that there was no way to escape the heat often brought us to the

36. In the Sahara, on the border line between a desert of black rocks coated with manganese and iron oxides and a desert of reddish sand

37. The desert locust, which imitates in form and coloring the oxide-coated stones that pave the ground in this part of the desert

verge of despair. The overheated air moved so violently that we seemed to be surrounded by a colorless ring of flame. The source of it all, the sun, could hardly be seen. The entire sky, from horizon to horizon, was filled with a glaring white, blinding brightness. We could remain in our observation posts only a short time and only if we first soaked our clothing in water.

The question that had become fixed in our minds, the one that we were determined to find an answer to, was: how could the broody sand grouse endure such heat? We had taken measurements and had discovered that the rocky desert floor rose to a temperature of 68° C. (155° F.) in the afternoon. How did the birds prevent their eggs from being roasted in the shell? We did not get anywhere by sheer reasoning. Only after I observed a pair of sand grouse behaving oddly did I hit on the right track. The birds were running across the rocky ground at an unusually fast pace. While the male, with raised head, scanned the area, the female was obviously looking for accumulations of sizable stones. When it found one such place, it quickly ran to the spot and with sidewise motions of its body scooped out a shallow basin.

After sitting quietly in this hollow for a while, it ran restlessly on. In the course of an hour it repeated the procedure with five different accumulations of stones. There could be no doubt that the female was searching for a suitable brooding site.

But why did none of these places suit her? As far as I could judge from my previous observations, all of them met the sand grouse's requirements for their brooding places. The color of the pebbles harmonized perfectly with the coloration of the birds, and there was an ample number of the "cover stones" that look enough like the birds to deceive cruising falcons. Yet the pair's persistent search showed me that some other factor—apparently a highly important one—must determine the choice of a nesting site. The insight came suddenly. It was as though a spark leaped between one part of my brain, in which my biological observations were collected, and another part, where my geological knowledge was stored. Each time the female lingered in the hollow it had scooped out, it must have been testing the temperature of the substratum. It was trying to find a "cool" spot in the midst of these hot surroundings. But how could there be cool spots in that monotonous wasteland of rock, and what might have caused them to arise?

From my geology, I knew that rocks of different densities heat up at different rates. Here there were no noticeable differences in the character of the surface for thousands of square miles. Yet my guess proved right. After all the chicks were hatched, I checked the substratum of the various sites where the birds had nested; I also checked the nesting places of the year before, which I had marked. I discovered that the substratum in this part of the desert was quite varied, despite the uniform appearance of the weathered surface. Some areas consisted of firm clay on packed sand, for instance. Alongside them would be areas composed of massive rock formations. But my most interesting discovery was that there were islands of very porous rock, 40 to 60 per cent of which consisted of bubblelike air-filled pockets (see Plate 28). The islands varied in size from a few square feet to one tract of about 250 acres. The nesting sites of the sand grouse were always on these islands of porous rock. The air-filled pockets make the rock a poor conductor of heat. Its temperature rises only to 45° C. (113° F.), while the denser rocks of the Sahara can reach temperatures as high as 70° C. (158° F.), or even, in extreme cases, more than 80° C. (176° F.).

Even though the islands are very small, the high temperature of an adjacent dense, massive rock cannot be transmitted to the "cool" island. The porous stone has the same effect as man-made insulation.

Once I knew what the sand grouse females were after, I was able to find the cool spots merely by feeling the ground with my hand, which had become quite sensitive to temperature variations. That was what the sand grouse females were doing when they scooped out their test hollows. Yet the hunt for a nesting place was no simple affair. It was not enough just to locate a cool spot. The spot also had to have cover stones. For birds that require such special conditions, the suitable nesting sites are very few.

This time I looked more thoroughly into the matter of the cover stones. It turned out that, in addition to providing visual protection for the brooding birds, they served another, quite different function. They were, as I have said, the last remnants of strata of rock hard and dense enough to have resisted weathering until now. Because of their density, they cooled off at night much faster than the porous rock on which they lay. Consequently, the small amount of moisture in the desert air would precipitate first onto these stones. In other words, they acted as a kind of dew accumulator. The same phenomenon can easily be observed in a moist climatic zone. If a material of high density, such as iron, is placed in the middle of a material of low density, such as wood, by morning so much dew will have precipitated on the iron that it will be dripping wet.

It seems to me conceivable that the moisture attracted by the cover stones is stored by the porous rock of the nesting site. The stored moisture evaporates in the course of the day, cooling nest and eggs. By day, of course, the cover stones have the disadvantage of absorbing too much heat. But apparently that disadvantage does not outweigh their numerous advantages. The high temperature is not transmitted to the porous rock on which the eggs lie. Also, the broody female sits far enough away from the cover stones so as not to be singed by them. Besides, the bird is able to survive outside the nesting site later on, in midsummer, when it is constantly exposed to an even hotter environment. During the brooding period its primary task is to protect the sensitive clutch of eggs from overheating.

The insulating effect of the porous rock and the presumed cooling effect of the evaporating dew just sufficed, according to my observations and temperature measurements, to assure the viability of the eggs. As the season advances, the atmospheric water vapor drops to the extreme of only one-twentieth the normal value in moister climates. This means that throughout the month of July, while some birds are still incubating their eggs, the quantity of dew at night is virtually zero. Yet the temperature at the brooding sites remained constant even when cooling by the evaporation of dew had ceased entirely. Clearly, other cooling factors were involved. I have already spoken of certain layers of rock beneath the surface of the desert in which large quantities of rain water are stored, water that fell from the sky thousands, if not millions, of years ago, when the continents were much wetter than they are now. Scientists have only begun to study this whole matter.

It is my guess that this ground water may play a large part in the cooling of the sand grouse nesting sites. Of course, the ground water has remained in place over long ages of time only because most of the water-bearing rocks are covered by a stratum of impermeable rock that prevents the evaporation of the fossil water.

But it is conceivable that the islands of porous rock on which the sand grouse situate their nesting places extend down, here and there, to a water-bearing stratum. The extraordinarily high rate of evaporation in the desert—in the Sahara a 20-foot column of water evaporates in a year—could draw the ground water out from under the islands of porous rock. The resultant cooling by evaporation would protect the clutch of eggs. The more the air temperature increases in midsummer, the faster the evaporation. Perhaps a balance is achieved between rising ground temperature and cooling by evaporation—

right at the surface of those islands of porous rock. That, at any rate, is my hypothesis. I hope to be able to find corroboration for it on a future expedition.

The combination of all these factors has led to the creation of a cool microclimate in the midst of searing surroundings. The sand grouse have been able to find these tiny climatic oases, and so they have managed to survive in one of the least livable parts of the earth's surface.

We decided to spend another five months in the Sahara the following year, 1970, in order to round out our observations on the sand grouse and, still more important, to make a documentary film about the most spectacular and significant aspects of sand grouse behavior.

Again, as in the previous year, watching the sand grouse provided us with countless new insights into the lives of these birds. There seemed to be no end to the discoveries to be made about this remarkable species.

In 1970, living conditions in the desert had actually worsened by comparison with the year before, which had been bad enough. Now even the majority of the spotted sand grouse seemed unable to cope. According to our observations, more and more of the adult birds simply skipped an entire brooding period because conditions were so harsh. We found very few brooding pairs, and at the height of the brooding season we often saw large flocks of nonbrooders. Apparently they had not found any other part of the Sahara where conditions were more favorable and were just waiting for better times.

It was hard to believe, but the more the spotted sand grouse were thrust back by the merciless desert, the more frequently individuals of a new species appeared to fill the void: the coroneted sand grouse (*Pterocles coronatus*). This species, whose survivability exceeded anything we had yet seen, turned up, along with blazing sandstorms from the interior of the Sahara, to nest in our study area.

Few Europeans have ever set eyes on the species, aside from a few stuffed birds in museum collections, and virtually nothing was known about it. The coroneted sand grouse's coloring is quite different from that of the spotted kind, whose ground color is medium brown, slightly sandy. The coroneted sand grouse has a much darker, reddish brown plumage with a distinct cast of purple. In addition, the male has a startlingly blue beak and, on either side of its head, two wide, brilliant blue stripes that meet in a large blue spot at the back of the neck.

Our experience with the spotted sand grouse had taught us that the birds were likely to nest where the color of the rock matched the color of their feathers. So we adopted a new technique in our search for the brooding places of the coroneted sand grouse. We did not go looking for the birds in the midst of vast areas of desert, as we had done with the spotted sand grouse. Instead, we searched for rocky regions whose colors corresponded to the highly specialized coloration of our subjects. There must be patches of desert, we reasoned, whose coloration contrasted with that of any other part of the desert, just as the coloration of the coroneted sand grouse contrasted with that of every other species of sand grouse.

We found those places. They were relatively small and consisted of coarse, weathered sandstone that was reddish brown to purplish in color. Even the blue marking on the male's head was there, in the form of small azure pebbles lying scattered among the sandstone rocks. Just as common were small white pebbles (see Plate 32). We learned later that the coroneted sand grouse chicks, during their first days of life, have strikingly whitish heads that match those pebbles in size as well as color. The males of the species supplied their chicks with water in much the same way as the spotted sand grouse did, but with greater assiduity. The coroneted sand grouse male brought water in its plumage twice a day, once in the morning and once in the afternoon. Even the technique of watering the chicks was adjusted to the conditions of the desert down to the last detail. When adopting its watering posture, whether morning or afternoon, the male took care to turn its back to the sun, so that the wet down could remain shaded, reducing the evaporation of the precious freight of water.

The chicks of the coroneted sand grouse displayed clear signs of competitive behavior—a result of the extreme living conditions. As the chicks grew, their need for water increased, of course. Standing under the male's belly, the little things would stretch their wings as a way of preventing their brothers and sisters from drinking.

Out of the multitude of life forms in the desert, I have selected the sand grouse for such detailed treatment here not only because my wife and I were able to explore their habits and habitat so thoroughly, but also because these birds provide a telling example of how living organisms respond to the challenge of the desert environment. As our planet slowly becomes a desert, all life on earth will have to face that challenge.

# The Camel—Marvel of the Desert

The camel probably deserves first mention among animals that have successfully adapted to the desert. For most people, this animal is the very symbol of survival in the desert, just as for most people sand dunes are the desert.

First I must explain when a camel can be called a camel. It has been my experience that the question of one hump or two is a touchy matter. If, for example, during a lecture on the Sahara, a picture of a caravan is shown and is referred to as a camel caravan, some amiable lady or gentleman in the audience will politely but firmly point out that it was actually a dromedary caravan, because the animals in the picture had only one hump. Real camels have two, this person will say, and, besides, they are found only in Asia. The objector is right as far as the definition of a dromedary goes. But in common speech the one-humped dromedary of the Sahara is generally referred to as a camel, perhaps because that is the shorter word. In fact, occasionally an entirely different two-humped creature is also called a camel. But to call a dromedary a camel is stretching the point no more than to refer to a cart horse and a race horse as horses. The one-humped and the two-humped species both belong to the same genus, *Camelus*. Zoologically, it is perfectly correct to refer to one-humped and two-humped camels.

The one-humped camel of the Sahara is not an animal that was recently captured and domesticated, as many people think. The only one-humped camels known today are purely domestic animals; nowhere in the world does this species still occur in the wild. Its wild ancestors, which were domesticated in Mesopotamia about 5,000 years ago, have long been extinct. The one-humped camel reached Egypt from Asia Minor about 4,000 years ago; but it spread farther westward, into the Sahara, relatively late, probably only a few decades before the beginning of the Christian Era.

There are, however, some wild specimens of the two-humped camel. Small herds of them roam the remotest portions of the Gobi Desert in Mongolia and China. Wild camels can also be met with, astonishingly enough, in America.

I made my first acquaintance with the American camels while traveling through the frightfully hot and dry Mojave Desert of California. In a parched and remote mountain valley I came upon the hoofprints of several horses, and only a short distance away I was struck by the sight of the highly characteristic prints of camels' feet. The ground had been trampled by them. From the number of prints, it was clear that a large herd had passed through. Like the prints of the horses, the camel prints looked quite fresh. The light-colored clay surface had obviously been moistened by a rain shower shortly before the animals came into the valley, and the hoofs had pressed perhaps an inch into the softened clay. Mud had been squeezed up along the outer edges of the hoofs so that a small bulge formed around each imprint. Among the prints innumerable tiny pockmarks could be seen, also surrounded by small bulges. The pockmarks were the prints of raindrops—the clay ground had not quite dried after the first shower when a subsequent brief rain left its impression on the still soft surface.

The prints of the camels were unusually small, only an inch or two in diameter. Full-grown, the camels could not have been any bigger than large dogs. Such tracks had been found only three times before. And yet, remarkable as it was to have encountered the extremely rare—and, what is more, fresh-looking—tracks of those tiny camels, there was no point in following the tracks in the hope of seeing one of the miniature beasts. The animals that left those tracks became extinct millions of years ago. So did the horses, whose prints also were tiny—the horses would have been no bigger than small ponies. And in adjacent strata I discovered the footprints of another long-extinct species: a mastodon. Skeletons of this early type of elephant have been found widely distributed in the rock strata of North America.

The prints I had found were fossil tracks. The animals had left them on the surface of clayey sediments that, over the course of millions of years, were deposited in a natural hollow. After the animals trod on the wet and softened clay, the clay dried out. Then the surface was covered by more clay deposits; in time thousands of thin layers of sediment covered the hoofprints. The sediment eventually hardened into sedimentary rock. One day that rock was uplifted by the movement of the earth's crust, and in the process it broke up. The weathering of the last few hundred thousand years removed layer after layer, until the stratum bearing the tracks was exposed.

I had been extremely lucky to come upon the hoofprints. Their extraordinarily well-preserved condition and the fact that even the tiny impact of the raindrops could be recognized after millions of years suggested that the stratum had been exposed by weathering only a comparatively short time ago. Within a few years, such a stratum, consisting of very soft, friable stone, would be destroyed. If I had not come upon the place exactly when I did, I might never have seen this evidence of the wanderings of ancient camels, horses, and elephants.

The tracks of the animals were not all in the same stratum, however. The folding of the mountain ranges that enclosed the valley had caused sediments of very different ages to be uplifted and to reach the surface of the earth again. The entire valley was filled with shattered sedimentary blocks, frequently piled every which way. Along the fractured edges of the blocks the individual layers could plainly be seen; often they were so thin that they might be compared to the pages of a closed book.

In the lowest and presumably the oldest layer (but folding can play tricks on the time sequences of geology), the traces of the dwarf horses and dwarf camels were close to each other in both space and time. The tracks of the elephants, on the other hand, were impressed into a high, relatively young stratum. The mastodons had tramped over the softened clay ground in the western part of North America at the end of the Tertiary period, about 2.5 million years ago.

The alert reader will at this point suspect a contradiction. Species of animals are indicators of climate; or, to put it differently, the appearance of a given species in a specific region permits us to make deductions about the climate. We also know that only certain species of animals can exist in certain climatic zones.

The elephants of our modern era require a damper, more humid climate, with lusher vegetation, than camels do. With a few reservations, that empirical knowledge can be applied to the ancient, extinct species. Although we cannot say that the dwarf camels of millions of years ago were authentic desert animals, we may assume that they and the dwarf horses lived in a dry, steppelike climate. The characteristic adaptations of present-day camels to desert life would certainly have begun to develop that long ago in the evolution of this ancient genus.

On the other hand, the early elephant lived only 2.5 million years ago in a much moister climate, which promoted a richer growth of vegetation. And here is where the contradiction seems to lie. The succession of the tracks shows that the climate on earth apparently was growing not drier but wetter. It is the same problem that was mentioned at the beginning of this chapter. In very recent geological times there were successive ice ages that, for brief intervals, interrupted the evolution and spread of deserts.

Horses and camels developed in North America some 30 million to 35 million years ago, at the same time as the first proboscidians, the primeval elephants, were evolving in distant Africa. In time the latter animals spread through what is now Europe and Asia. Then, as the climate gradually cooled, more and more rain was bound up in the form of snow and ice. Abundant rain no longer flowed down the rivers and back into the sea, but was stored on the land as ice. The water level of the oceans dropped dozens of feet. Narrow, shallow straits that had previously divided some of the land masses became land bridges. The continents of North America and Asia, now separated by the shallow Bering Strait, were united by one of those land bridges. And about 2.5 million yeas ago elephants migrated to North America across that bridge. Horses and camels took the reverse route; those animals, which had meanwhile evolved into much larger creatures and had developed many species, crossed the Bering land bridge to Asia. It is not impossible that elephants and camels met in Alaska or in eastern Siberia. It would have been a strange sight—a herd of elephants migrating into Alaska from the Far East meeting a westward-bound herd of camels.

Two small North American species of camel, the guanaco and the vicuña, migrated to South America by way of the Central American land bridge, which had been formed by volcanic activity. Both species still live in the wild in arid regions of South America. Later the domesticated llama was bred from the guanaco and the alpaca from the delicate vicuña.

The ice age spelled the end for all the horses and camels of North America. The newly immigrated elephants also disappeared. Only in historically recent times, between the sixteenth and the eighteenth centuries, were horses and camels brought back to North America, by the European settlers. Both families of animal had spread throughout the rest of the earth, first of their own accord and later with the help of man, before they met again on the continent where they had originated. Without the horse, human settlement in the Americas can hardly be imagined. The settlers, in their surge westward, were completely dependent on the labor power and endurance of the horse, which very rapidly adapted to its new-old biosphere.

In contrast to the horse, the camel could not readjust to its former home. Except in zoos, camels can be found nowhere in North America today. Why they failed to thrive in the deserts of western North America is not clear. The camels that migrated from America 2.5 million years ago probably were still, like the horses, genuine animals of the steppes. Perhaps it was not until they reached Asia that the camels underwent the evolution of making an extreme adaptation to desert life, and the process was certainly carried further in

Africa, after man introduced the one-humped camel into the Sahara.

The more extreme the biosphere of a given species, the more specialized are the adaptations that the species evolves. That generalization applies especially to adaptations to desert life. There is the additional factor that one desert is not like any other. A desert animal cannot simply be shifted from one desert to another. But one forest is not like another, either, and it would be impossible to transfer a bird of the tropical jungle to a northern beech forest.

The horses that were brought to North America by the settlers quickly and easily adjusted to life on the American prairies but not in the deserts, for they had originated on this continent as prairie animals. The characteristics evolved millions of years earlier were still largely in harmony with the land; it was as if the genetic heritage of the horses still "remembered" their old homeland. A few attempts were made to reintroduce camels, but they, on the contrary, had acquired new adaptations "abroad," in the spacious sandy deserts of Asia and Africa. In North America, they encountered an entirely different type of desert, consisting of innumerable jagged mountain ranges and rocky highland plateaus cut by canyons 6,500 feet deep. Any flat land was covered with sharp-edged stones that cut the camels' soft feet, which had adapted to pacing over sand. In the Sahara, camels are very reluctant to walk on rocky ground, let alone travel through mountains. But sandy deserts are rare in North America. The continent's regions of sand dunes are relatively small in extent; the huge river system of the Colorado transports most of the sand produced by weathering all the way to the Pacific Ocean.

Further support for our thesis that camels thrive best in flat, spacious, and sandy deserts is the fact that the camels brought to Australia by the first settlers became one of the most successful species ever introduced on that continent. The Australian desert shows little variation in altitude, and that seems to agree with the camel's physiology.

The camel is the largest mammal that has succeeded in adapting to desert life. (The two South American species of small camels are an exception.) The two large species are so alike in their physique and habits that whatever is said about the one can with slight modification be applied to the other.

We have seen some animals of the desert try to protect themselves from the heat of their surroundings, try to cool their bodies in order to survive in the fiery sunlight. Many species pant and sweat, reducing body temperature by the evaporation of water. Other species hold their vital functions to a minimum by day in order to avoid producing body heat; they live nocturnally. The sand grouse have discovered the most ingenious methods of cooling themselves. The camel, however, has not adopted a nocturnal mode of life. Nor can it, because of its size, creep away and hide. Even in the hot afternoon, when all other living creatures, including men, have taken refuge in the shade of small bushes or projecting rocks, camels can be seen imperturbably marching along.

The camel is a wonder of nature. Every feature of the animal is contrived to assure survival in the desert. If we observe a camel from a distance, three characteristics at once strike the eye. First, there are the long legs, unusual even for an animal of its size; second, the large body with the hump or humps; third, the long neck with the head perched on top. On the basis of these external characteristics, we can divide the camel into three "floors."

The lower floor includes the legs and ends with the underside of the belly. The long legs give the animal its running speed and endurance. The camel's life is dependent on those two factors. Whenever possible, it must be able to escape its enemies by swift flight. At least, that was true in the remote past; today's domesticated camels have no natural enemies. (Little is known about the extent to which natural enemies affect the lives of the last wild camels, in the Gobi Desert, or whether enemies exert any significant selection pressure on the animals.) The second advantage of the long legs, great endurance in running, is more valuable than ever to the camel. In particular, the one-humped camel of the Sahara must cover longer and longer distances to find food, which is steadily becoming sparser. And the long legs serve still another purpose. The hot layer of air that forms above the ground in the summer ends at about the level of the underside of an adult camel's belly. The long legs raise the body, in which all the vital organs are situated, into a slightly cooler layer of air.

Finally, the remarkably shaped feet also belong to the lower floor. They are strikingly large and rounded; in an adult animal they can attain the size of a dinner plate. Like snowshoes, which keep the walker from sinking into the snow, the huge feet of the camel permit it to trot over soft sand without sinking in.

According to my observations, the people of some nomadic tribes of the Sahara have larger and broader feet than Europeans do. Might that phenomenon be a first sign of a human physical adaptation to desert life?

The camel's second, or middle, floor includes the trunk. Most of the physiological processes that make it possible for the animal to survive in the desert take place on this floor.

Warm-blooded animals that live in temperate or cold climatic zones maintain a constant temperature gradient between the warm body and the cooler environment. The reverse is true for the camel. The temperature of the air, in the hot sunlight, is usually higher than that of its body. If the high exterior temperature were to be transmitted to the body, the camel would die. It partly prevents this with the very shape of the body, which might be compared to a lens standing on edge. Remember that in the hottest summer months the sun's rays at noon fall vertically. The lens shape keeps those intensely hot rays from striking perpendicularly over a large part of the body and heating it unduly. It is during the afternoon hours, when the sun stands at an angle of 20 to 30 degrees above the horizon, that camels do their utmost to avoid exposing their bodies broadside to the sun's rays. By their peculiar shape and by their conduct, the animals insure that most of their body surface lies in full or partial shade during the hottest time of day. That makes it possible for them to radiate away body heat.

The camel's body has other means of forestalling dangerous overheating. Animals that live in cooler climates usually have a layer of fat evenly distributed around the body just under the skin. This subcutaneous fat serves chiefly to keep heat in by preventing its radiation from the body. The camel has no such layer. It stores almost its entire supply of fat in its hump or humps. The animal's body tissues, uninsulated by fat, are directly beneath the skin, where they can radiate away body heat.

Another desert mammal, the small fat-tailed mouse (actually a gerbil), utilizes the same method to get rid of heat by radiation. As the name indicates, the greater part of its body fat is stored in its stout, clublike tail.

The camel's hump has inspired all sorts of fantastic stories. For example, it is a common belief that camels can store large amounts of water for long periods in their humps and in their stomachs. The Roman geographer Pliny reported in his *Historia naturalis* that the camel had a water-storing organ in its body. Although dissection failed to reveal any such organ, the idea remained current up to a few years ago. Countless stories have been related of people saving their own lives by slaughtering their faithful camels and drinking the water from their humps or stomachs. I shall never forget a picture I came upon some time ago in a book on desert adventures. A lively drawing showed a desert traveler in an advanced state of distress simply stabbing his beloved camel in the hump. Where the knife blade had penetrated, out sprang a jet of pure water. Until recently, even zoology textbooks referred to such a method of water storage.

It is a fact, however, that camels can survive without drinking or eating for up to two weeks and yet retain the strength to cover several hundred miles. That capacity, for which camels are famous, represents one of the most interesting adaptations any animal has evolved for survival under extreme conditions. Because of it, camels are able to traverse the great distances between widely scattered oases, where food and water are to be found. Indeed, it is the stamina of the camel that has made it possible for men to cross and to live in the deserts of Asia and Africa.

The camel can drink up to 250 pints of water at one time. The water is temporarily stored in the stomach, which has a capacity of 525 pints. Within a very short time, part of the water is evenly distributed through the animal's tissues. The rest is absorbed by the red blood corpuscles, which in the camel have a unique property. They can expand to 240 times their original volume, and thus are able to store enormous supplies of water.

But every vital process is dependent on a constant input of energy. In order to create energy by metabolism, living organisms must take in nourishment at short intervals, in addition to oxygen and water. How do camels solve the problem of nourishment during their long fasts? Under normal conditions, they neglect no opportunity to gobble up large amounts of food. They feed chiefly on camel's thorn, a bush whose prickly shoots and leaves contain a great deal of water. If an ample supply of this food is available, a camel will quickly acquire an impressively fatty hump. In the one-humped camel, the hump can weigh up to 30 pounds; the two humps of the Bactrian camel can reach a weight of 50 pounds. As the hump grows, a good part of the water the camel has drunk is chemically bound and incorporated into the fat of the hump.

During periods when the desert is parched, or on long treks across regions without water or food, the camels depend on the supply of fat in the hump. Their water needs are met by the water stored in the red blood corpuscles, in the fat of the hump, and in the body tissues. In addition, a great deal of hydrogen is produced during the metabolism of the stored fat, and it unites with oxygen taken in by respiration to form water. So the camel has developed the ability to water itself for a long period of time. The amount chemically produced by the metabolism of fat is considerable. When half a pound of fat is burned, about a pint of water is produced. Camels that have gone without water or food for a protracted period often lose up to 30 per cent of their body weight. They look emaciated and weak; the ordinarily plump hump is slack and shrunken. But the animals are not sick. When, after a long, hard journey, they at last reach a patch of vegetation and a water hole, they rapidly devour great quantities of food to build up their supply of fat. To replace the water lost from the fat in the hump and to pump up the red blood corpuscles, they will drink more than 200 pints of water in ten minutes.

But consuming such great amounts of water is not the same as building up a store. Rather, the camel is merely reconstituting the normal condition of its body. It does not drink with an eye to the future, but to make up for the past.

It is important that the camel lose as little of its vital water as possible. Most mammals, including men, need a considerable amount of water in order to excrete the urine produced by the kidneys. Whereas human urine must consist of about 95 per cent water before it will leave the body, some desert mammals, including the camel, have developed methods for excreting urine in a considerably more concentrated form.

The most amazing characteristic of camels—the one that truly sets them apart from other warm-blooded animals—is that they normally neither pant nor sweat in order to cool their bodies, and so they can conserve the most precious substance in the desert, their body fluids. When the summer temperature rises above 50° C. (122° F.) in the shade, the camel can radiate no more body heat, even from the parts of the body that lie in the shade. The cooling effect that results from the shape of the body is no longer enough to guarantee a constant body temperature. But the camel has a further characteristic, one that is unique, so far as we know, among mammals and birds. Without any impairment of its physical capabilities, the camel can withstand an increase in the temperature of its body—and of its blood as well—of 9° C. (16° F.) above normal. Applied to human beings, that would mean we could do heavy physical labor with a body temperature of 46° C. (115° F.).

Not until the external temperature reaches a height that is extraordinary even for this uniquely adaptable animal does the camel begin to lose water to the atmosphere by evaporation, that is, by sweating. But that happens only at a point at which other mammals, including man, already would have lost most of their body fluids from sweating heavily. They would be dead before the camel began to sweat. And even sweating does not imperil the camel, because the amount of water in its blood does not drop to a dangerous level.

The animal's third, or top, floor consists of the tip of the hump and the head. We have already noted the advantage provided by that head perched on its long neck and reaching above the probable upper limit of most sandstorms, allowing the camel to breathe air relatively free of sand.

The millennia of experience with life in the desert that camels have accumulated are expressed not only in physiological traits but also in distinct patterns of behavior. The combination has enabled these animals to survive. An ethnologist* once remarked lightly that the first prize for the conquest of the Sahara must be given to the camel, since it had been the first vehicle to meet desert conditions. He listed its specifications: four-wheel drive, three shifts, a large fuel capacity with reserve tank, special dune-buggy tires, an extra-strong front end, and a visible fuel gauge in the form of a hump.

In this chapter I have given some account of how various forms of life manage to survive in the hostile environment of the desert. From these examples, it is clear that living organisms are engaged in a continuous reciprocal interaction with their environment. The environment includes a number of life forms, but it is chiefly shaped by inanimate nature: the interplay of the elements, the energy of sunlight, the gases of the atmosphere, the rock of the earth's crust, and the built-in order of events to which all matter is subject—its observable succession in what we call time. Two of these factors are seen with signal clarity in the desert: the glaring extraterrestrial light of the sun and the terrestrial cycle of the rocks—weathering, denudation and rearrangement, transport, folding, and, finally, the formation of ever higher mountains.

These forces, and the laws that govern them, shape the earth's history. The living creatures of the desert are only acting out the briefest of scenarios. Can we truly say that they have adapted to their extreme environment? Might it not, rather, be said that their adaptations, even the organisms themselves, are merely products of the coercive forces of the cosmos? In the unequal struggle, can any organism or species permanently overcome its inanimate environment?

I have already pointed out that for man the answer is no. There are

* G. Gerster, in his book *Sahara*, published in Berlin in 1959.

limits to the adaptability of living things. Plants and animals can make truly astonishing adjustments to desert conditions, but as the climate on the continents becomes steadily more arid a point will be reached at which not even the cleverest adaptations will avail. Then life will retreat entirely, leaving the desert waste and void, as it was in the beginning. The retreat does not take place overnight; rather, it proceeds slowly over many stages. Often the struggle of living organisms against aridity goes on for millions of years.

# 5 / The Retreat of Life

## A Cold World

In spite of the big snowshoes we had strapped to our boots to keep us from sinking into the waist-high powder snow, we made laborious progress. Before us stretched an endless-seeming white wilderness in which the two dots that were our bodies appeared to be fixed in one spot. Step by step we fought against an icy arctic wind that blew in our faces from due north.

Yesterday a blizzard had come up unexpectedly. Within an hour a mass of black clouds had climbed into the northern sky. Its leading edge stood out like a line drawn with a ruler on the clear, cold blue sky. Under that crushing black wall, which piled higher and higher, glowing white whorls of blowing snow could be seen. Soon the blizzard was upon us. I had never before seen such masses of snow in the air. The sun was completely hidden, and a soft twilight seemed to swallow all outlines.

By the next morning the weather had improved, and we set out on our expedition. The black bank of clouds had vanished. The sky reappeared in an almost colorless gray light. The red disk of the sun, low on the southern horizon, produced a faint reddish shimmer in the sky. Its cold light was reflected by the innumerable tiny ice crystals still drifting about in the air. It was the first time I had seen a landscape radiate such intense cold. Even the sunlight felt icy. The sharp arctic wind did not allow the daytime temperature to rise above $-20°$ C. ($4°$ F.). We wore heavy winter clothing, and our heads were protected by thick hoods with slits for eyes and mouth. Since the snow was still drifting thickly just above ground level, our progress was considerably slowed despite the clear view. The endless carpet of snow flowing along the ground made us lose all sense of distance, and the high speed at which the snow moved produced feelings of giddiness. The snow blotted out changes in ground level, so that with each step we had the nasty feeling that we might be plunging into a hidden crevasse. In some places the surface was so hard and rocky that we were able to unbuckle our showshoes and make faster progress. But then we would step into stretches of waist-high snowdrifts that had formed in the lee of low mounds. These we could cross only with the aid of the snowshoes.

Our goal was a desert whose sand hills, more than 800 feet high, are among the highest desert dunes on our planet. We were almost upon it now. The dunes were overtopped by a snow-capped and ice-covered massif 15,000 feet high that loomed up directly behind the sandy desert. When at last, considerably later than planned, we reached the foot of the first dunes, we stopped to reckon how much time we had left for our studies before we would have to start the tramp back. We had to arrive at our heated base by sunset. After sunset, the temperature here could drop to $-55°$ C. ($-67°$ F.).

Those mighty sand dunes were an astounding sight. Over the surface of the grayish yellow billows of sand drifted dense streaks of snow. The scene reminded us of an ocean churned up by a storm in which the wind drives long bands of white foam over the surface of dark, gray black, gigantic waves. Here and there along the slopes of the dunes the wind had scooped out and rearranged the sand to such an extent that the inner structure of the dunes was visible. What we discovered was fascinating. The cross-stratifications of the sand made it possible to recognize the prevailing wind directions, but in addition they had the peculiarity—surely unique for sand dunes in a desert—that they consisted of alternating layers of sand and snow. It was easy to read off some of the climatic factors that had led to the formation of this singular desert. The snow had been blown in and deposited chiefly by winds from a northerly direction, while the sand had been transported by southerly winds or else had shifted from one stratum to another inside the dune.

That astonishing sandy desert with an arctic climate is located in the southern part of the United States, in the states of Colorado and New Mexico. It is a part of the great North American relief desert, and in temperature it ranks with the most extreme regions of the globe. In summer the temperature can rise higher than $40°$ C. ($104°$ F.). That is, there is an annual temperature differential of nearly $100°$ C. (or more than $170°$ F.). Such an extraordinary fluctuation is not caused solely by the continental climate and the great altitude of the desert; a principal reason is the north-south orientation of the mountains that surround all the North American deserts. Even as those mountains produce desert conditions by blocking off the moist air masses from the oceans, they allow an unhindered exchange between the cold, dry air masses from the Arctic and the hot but also dry air from the subtropical deserts of Mexico. During the winter months an arctic snowstorm can, in two or three days, pass from northern Canada down over these deserts of the south, a distance of 2,500 miles. The storms bring not only great cold but also, in the form of snow, the larger part of the sparse annual precipitation. In the summer, the prevailing winds from the south are hot and laden with sand and dust. As a result, parts of the high relief deserts of North America are both the hottest and the coldest deserts on earth.

The remarkable stratification of sand, snow, sand, snow, and so on, is found only during the winter months. In that season southerly winds—which do not raise the temperature above the freezing point —frequently blow sand on the snow that was earlier deposited by a blizzard. When the hot winds recur with the onset of spring, the accumulated layers of snow melt.

Along with the great aridity, the extreme temperature fluctuations pose an enormous problem for the life forms that attempt to survive in these deserts. Once more, the desert birds would appear to have an advantage; like the many species of birds that nest in the Arctic during the summer, they can migrate when the icy winter descends. Insectivorous birds, in particular, would seem to have no other choice, since their food is simply not available in the winter.

One of the most common insect-eating birds of the southwestern deserts is the poor-will (*Phalaenoptilus nuttalli*), whose name, like that of the whippoorwill, is derived from its cry. It is about as large as a blackbird and is a member of the family of nightjars, or goat-suckers, which bear the scientific name *Caprimulgidae* and are dis-tributed world-wide. The poor-will can be observed especially at twilight in spring and summer; it is a nocturnal bird and does not set out to seek its food until evening. Because of its rounded wings, it somewhat resembles a giant moth, and its irregular, jerky movements in flight have been compared to those of a bat. The poor-will hunts nocturnal insects, which it catches on the wing, with its huge, funnel-shaped beak and gullet wide open.

In the desert, the poor-will nests on the ground, and, like all birds of similar habit, it has plumage that is protectively colored to har-monize perfectly with its surroundings. Unlike the rather monotone Sahara, the relief deserts of North America often display a wide variety of rock and soil colors within a small area. That is due to their geological structure of numerous mountain chains, fissured plateaus, and deep canyons cut by rivers. As might be expected, the poor-will has many color variations. There are reddish populations that nest in the red sandstone deserts of the Colorado Plateau, buff-colored birds that prefer sandy sections, and gray birds that brood among the granite mountains of the Mojave Desert. Still another type, which broods in the cactus desert of southern Arizona and in Mexico, has dark, barklike markings on its plumage that exactly match the color of the dead cacti among which the bird nests.

When insects disappear with the onset of the icy desert winter, the poor-will vanishes with them. It was believed that it moved into the warm south, like other migratory birds, but its winter quarters had never been discovered. On December 29, 1946, the American zoologist Edmund Jäger was with a group of students on an expedition in the Chuckawalla Mountains. He had no suspicion that on that cold winter day he was about to make a sensational biological discovery.

These granite mountains within the Mojave Desert of southern California form one of the most impressive landscapes on earth. The granite has been weathered into extraordinarily bizarre forms because of the extreme fluctuations in temperature. Every shape of rock imaginable seems to be present here. Over millions of years the cliffs have been split and split again, and the sandblasting of the desert wind has widened the clefts and polished individual fragments and blocks of rock into fantastic shapes. The mountains are made up of thousands of huge spheres and rhomboidal blocks. They look as though giants had been tossing around massive building blocks (see Plate 33). When I first clambered around in these mountains, in 1973, I was constantly troubled by the feeling that the huge heaps might collapse at any moment, like a pyramid built of balls. The extended plateaus that stretch between the higher ranges are covered with thousands of monumental stone sculptures that seem to be the work of countless generations of sculptors, although they are really the last remains of mountains worn away by weathering. Among the rocks, solitary or in groups, grows a species of yucca, a member of the lily family, that stands 30 feet high. These Joshua trees, with their bizarre, armlike branches, add their note toward making the whole mountain range a surrealistic dream landscape that seems to have sprung from the imagination of a Salvador Dali.

As Edmund Jäger and his student group were passing through a small, narrow canyon, his practiced eye spotted something curious on the wall of a cliff. In one of the innumerable shallow hollows no bigger than a hand, which weathering had worn in the rock, he thought he noticed a hint of something. When he got closer he saw a bird sitting in the hollow. It was a poor-will, its spotted gray plumage harmonizing perfectly with the crystalline structure of the gray granite. The hollow was just big enough to hold the bird. The group watched it for ten minutes without its giving the slightest sign of life, and then Jäger stretched out his hand to touch it. Even when he ran his hand down its back several times, the poor-will gave no reaction to suggest that there was any life left in it.

The group tramped on. When they passed through the small canyon again two hours later, on their way back, the bird was still in the hollow, its position unchanged. This time Jäger reached in and took it out. It was unusually light and seemed to be dried out. Its feet and its eyelids, on which a bird's temperature can be felt, were quite cold. Evidently the poor-will had not flown south, for some reason, had been caught by a sudden cold wave, and had frozen to death in the little niche where it sought shelter. Jäger replaced the bird in the hollow, and the group continued on their way.

Ten days later, Jäger brought a friend, Professor Lloyd Smith, into the Chuckawalla Mountains to show him the unusual find of the poor-will. The bird was still in the same place, its position unchanged. But when one of the men took it out of the niche and was holding it to examine it, something unexpected happened. The bird's beak opened, and the poor-will uttered a few very high-pitched, whimpering, squeaking notes. Both men also thought they saw its eyes open briefly. Suddenly the poor-will raised its wings; but there was something oddly lifeless and mechanical about the movement. And even the peculiar sounds might have been caused merely by air that somehow was pressed out of the lifeless body in their handling of it. Jäger cautiously tried to fold the wings back into their previous position, but he could not. When he released them, they snapped back into the lifted position like steel springs. By using force, he finally managed to settle them so that they lay flat against the body. Then the bird was replaced in its hollow and left as an object to show future student expeditions. It no longer looked very good. The plumage had been tousled by all the handling, and the wings stood at an unnatural angle.

When the scientists returned two hours later, they saw by this unmistakable posture that nothing had changed. They decided to take a photograph, however, and took the bird in hand once more. At that it promptly flew up and away. The two friends could scarcely believe their eyes as they watched the poor-will, wings beating rapidly, fly adroitly up along the cliff wall and find another hollow in the rock, far beyond their reach.

How was this possible? The two men, both scientists, did not believe in miracles, and immediately they asked themselves whether they had been dealing with a bird in hibernation that had gradually awakened—disturbed by their handling—and finally had flown away.

Hibernation is a fascinating adjustment that has been developed chiefly by mammals of cool or polar climatic zones. Unlike birds, most mammals are not able to migrate when their food supply runs out with the onset of winter. Instead, many mammals retire from active life for a while; they go into their dens or burrows and fall into a winter sleep. During this period their metabolism is sharply reduced. All their vital functions are so throttled down that they need to consume only small quantities of the store of fat they have accumulated from their summer feeding. Mammals in hibernation have been observed to breathe only once every five minutes, and the normal rate of 300 heartbeats a minute declines to just 5 beats. The slowing of metabolism and other vital functions means that the body temperature drops rapidly, in some species to a few degrees above freezing. An animal in hibernation has only its pilot light going, so to speak.

But hibernation in birds had never been observed before. In comparison with mammals, birds are far more active, their metabolism much faster. Some birds have to take in several times their body weight in food every day to survive. How could birds ever reduce their vital functions enough that they could hibernate? It seemed impossible.

At the end of November 1947, almost a year later, Jäger had the great good luck to find the poor-will again, in the same cliff hollow. This time, he was able to conduct careful studies to verify the hypothesis that a bird—and, what was more, a desert bird—could indeed hibernate. First he measured the bird's temperature at the anus. The thermometer read 18° C. (64° F.). That was higher than the temperature of the surroundings, and therefore was evidence that the bird was still alive and was actually in hibernation. The bird, whose normal temperature is about 42° C. (108° F.), had lowered its body temperature by about 24° C. (44° F.).

A stethoscope applied to the poor-will's breast told nothing; the heart was beating so slowly and feebly that it could not be heard. A cold metal mirror held in front of the bird's nostrils did not mist over, as it would have if there had been any noticeable breathing. Jäger tried one more experiment to see whether the bird was alive. He opened one eyelid and flashed a penlight into the pupil of the eye. There was no observable reaction. The bird did not even try to close its eyelid. The only indication that it was not dead was the measured body temperature.

The bird was weighed at intervals of two weeks, and those measurements showed that the weight was slowly diminishing: 45.61 grams, 45.58 grams, 45.50 grams, 44.56 grams. The figures also demonstrated how very slowly the hibernating bird was consuming its reserves of fat. When the winter was over and the poor-will ended its three months of deathlike rigidity, it had consumed only 6 grams of fat.

Another interesting finding was that the poor-will chose the same hollow in the same cliff for its winter sleep in the following two years, 1948 and 1949.

After Jäger's account of his sensational observations was published, he received many letters reporting similar phenomena. A number of readers wrote of having found a "dead" poor-will and taken it home to show to their families; when the bird had lain for a while in a warm automobile or house, it had suddenly revived, as if by a miracle, and flown away. Even before the white man came to America the Hopi Indians of Arizona knew the poor-will's secret. They called the bird *hoelchko,* or "sleeper."

But why did the poor-will evolve the complicated bodily functions that are necessary for hibernation? Why did it not simply migrate to the south like other birds?

In the first place, the long journeys of migratory birds are not so simple as they might seem. On their way to and from their winter quarters, birds are exposed to countless perils, and casualties are high. The principal reason for the poor-will's hibernation, however, is not the dangers of migration. Migratory birds usually come from regions in which the change of seasons follows a rhythm that has been fixed for long geological ages, and, of course, fixed seasons are the norm for most of the continent of North America. But in many desert areas winter follows a highly irregular and unpredictable course. Sometimes winter can descend on the desert so abruptly, in the form of a blizzard, that the poor-will would not have time to flee. The bird would freeze or starve to death if it had not developed hibernation. There is the additional factor that because of its specialized desert coloration it would have difficulty overwintering in quarters outside the desert, where it would not be camouflaged.

The poor-will's hibernation has a unique aspect that protects the bird against short-term changes in the weather. Mammals require days and weeks to slow down their metabolic functions gradually before they fall into hibernation. The poor-will, however, can do so in a very short time. The transition from the active state to hibernation is so rapid that, for instance, a completely active poor-will can be placed in a refrigerator without suffering any harm. It at once drops into its winter sleep. The experiment also works in reverse. The bird can react to extreme fluctuations of temperature, no matter how brief or irregular, with the utmost flexibility—and just such fluctuations are characteristic of a relief desert. The cold weather of the desert winter is frequently interrupted for short periods by billows of warm air, caused by the rapid exchange of air masses resulting from the north-south orientation of the mountain ranges. The poor-will can use the periods of favorable weather to wake up and feed up.

## The Estivation of Mammals

Perhaps the best known of the desert mammals are the jerboas. These small rodents, which inhabit the deserts of Asia, Africa, and North America, are characterized by enormously long hind legs. They move with kangaroolike leaps and can attain incredible speeds. In individual leaps of as much as 6½ feet, they can cover 20 feet in a second. The human eye is hardly able to follow the movement of the hind legs. With the aid of its long balancing tail, the jerboa can change direction by almost 90 degrees while running at top speed. Anyone who has the good luck to observe these small nocturnal creatures on a bright moonlit night will gain the impression that they are propelled by soundless jet motors and are flying in wavelike movements over the surface of the desert.

Interestingly enough, the jerboas of the Asian and African deserts differ very little from the kangaroo rats of the North American deserts, even though the animals belong to different families, are only distantly related, and have evolved independently of each other. Adaptation to desert conditions has imposed the same form of locomotion on all these small rodents. And similar locomotion implies similar body construction, a largely similar bone structure, and a similar appearance. All the small desert rodents must be able to move as swiftly as possible, with the least exertion, over rough terrain where many enemies may be lurking.

The jerboas have a number of other characteristics that have evolved especially for life in the desert. Although their main food is dry seeds, they are one of the few species of mammals that never need to drink. They get their fluids almost entirely from the water of oxidation, that is, water released during their metabolic processes. Experiments have shown that a jerboa can produce 54 grams of water from 100 grams of seeds. Other animals, including human beings, also produce large amounts of water during metabolism, but much of it is excreted as urine or is exhaled, along with carbon dioxide, in the form of water vapor.

The jerboas are able to retain nearly all the water of oxidation. Purely nocturnal animals, they spend the entire day in underground burrows, in which the insulating layer of soil prevents the temperature, even on hot days, from rising above 20° C. (68° F.), and so they do not need to use water to regulate their internal heat. They do not sweat, and, furthermore, they have reduced to a minimum the loss of water in bodily excretions. Their urine is highly concentrated and sometimes contains twice as much salt as sea water does. Their feces, too, are completely dry and hard. Moreover, they exhale much less water vapor from their lungs than do other mammals, and they manage to make what they do exhale useful to them. While they are sleeping in their cool underground burrows by day, the entrance is closed with a plug of soil. The exhaled water vapor cannot escape and produces a humid microclimate in the burrow. There, the greater part of the water vapor is absorbed by seeds that the animals have stored against periods of dearth; the seeds are kept in a niche in the wall of the "bedroom." The sleeping jerboa always lies so that its nostrils are close to the seed supply, and the seeds, acting as a filter, extract the last drop of moisture from the air. In times of need, when the animal must eat the stored food, the water absorbed by the seeds is utilized again. In effect, the burrows of the jerboas contain a kind of closed system of water circulation, along with their food supply.

Despite such clever adaptations to life in the desert, these small rodents, unlike the sand grouse or camels, are not fully able to cope with the hardships of their environment the year round. During the winter they temporarily go into hibernation, like the poor-will. What is more, many small desert rodents go to sleep during hot, dry periods. This summer sleep is called estivation, and it probably represents a further development of hibernation. When, during the hottest season of the year, all the plants are withered and the last seeds have been carried away by the wind, the desert rodents throttle down their metabolism to a minimum and take to their burrows for months of sleep. That is how they survive the deadly season of greatest heat.

Kangaroo rats are able to practice both forms of long sleep: they can hibernate and estivate, too. Experiments with these animals have demonstrated that the kangaroo rat can be induced to hibernate by a drop in the temperature of its surroundings to the freezing point. Breathing almost stops, and there is hardly any consumption of its fat reserves. The animal needs only a minimum amount of energy to survive. Its body temperature, normally 39° C. (102° F.) in the active state, drops to 6° C (43° F.). But to throw the kangaroo rat into estivation, the experiments showed, it was not enough merely to raise the air temperature. It was also necessary to withhold food from the animal. Under that double pressure it would fall into summer sleep. The body temperature during estivation, in contrast to what it is during hibernation, drops to between 15° and 20° C. (59° to 68° F.).

Some species awaken from estivation now and then to eat some of their stored, moist food. That behavior probably serves to maintain the closed circulation of water in the sleeping chamber.

While the poor-will suspends its active life just during the winter months, the kangaroo rat, which lives in climatically extreme relief deserts, carries on its vital activities for only brief periods of the year. This animal has taken a considerable step in the direction of a retreat from life under desert conditions.

## Buried Alive

Autumn in the southern part of the American desert.

Our site lay in the midst of the wide dried-up basin of a geologically ancient lake. For shade we had stretched a canvas on four poles. Even so, the thermometer under the canvas read higher than 40° C. (104° F.). The air seemed to be afire; the ground was covered with a layer of white salt crystals. The result was that everything around us was bathed in a blinding glare whose source could not be perceived. The whole world, including the ground, seemed to consist of white light. We protected our eyes with very dark sunglasses. Early in the morning I had begun to work away with a pick at the hot, stone-hard soil. Every blow loosened at most a lump the size of my fist. We could only work mornings and evenings.

We were engaged in digging out a living organism—or, more precisely, hacking it out of the ground. It was an organism that very few people have ever seen. Residents who claimed to have heard it during the silent desert nights had told us that it baaed like a sheep. What we were searching for was not a sheep, but a creature bearing the scientific name of *Scaphiopus couchi*. As we were to learn later, our informants were not ribbing us. The animal's call actually did sound much like the baaing of a sheep.

The whole affair had begun several months before, in midsummer. At the time, we had set out to observe fish in the desert—not sand skinks, but real fish, which I shall come back to in a later chapter. Although finding a *Scaphiopus couchi* would, of course, have interested us enormously, there was little point in undertaking a systematic search for it. The chances of finding a specimen were virtually nil. Then a detour in the road unexpectedly came to our aid. From the detour road—or, rather, a track that we normally would never have driven on—we discovered a small pool in the midst of the desert. As we were told later, until recently not a drop of rain had fallen in these parts for more than three years. But the traces of a very recent rain on the surface of the ground indicated plainly that the pool had been left by one of the irregular downpours that occur no more than once every few years during the violent summer thunderstorms that are characteristic of this part of the desert.

Since the low-lying desert areas in the extreme southern part of Arizona and California have a high temperature throughout most of the year, the seasonal rhythm of animal and plant life here depends chiefly on the rains. If, in a given year, there are enough summer thunderstorms with subsequent cloudbursts, August and September will be a lush "spring." A dry winter will follow, but the temperature remains summery. In February there may be more rainfalls; these tend to occur more regularly than the summer rains, but they bring less precipitation. After the winter rains there is a brief spring, followed by another dry period, lasting until the season of the summer rainfalls. In other words, in this part of the continent the seasons appear several times in a year—if we measure "seasons" by precipitation and the activities of life.

In the pool that had collected in the lowest part of the basin we discovered several specimens of *Scaphiopus couchi*—the spadefoot toad. Toads are amphibians; they are absolutely dependent on the presence of water in their biosphere because their larvae, the tadpoles, cannot develop except in water. One hardly expects to find that they have survived in the desert, of all places. The amphibians, it will be recalled, were the first vertebrates to come out of the sea, some 350 million years ago, and settle on the protocontinents, when they were much flatter and wetter than they are today. In the highly developed relief desert of western North America, the amphibians should logically have been the first class of animals to be forced out. And, although our observations would seem to indicate the contrary, they are indeed being forced out.

Where had the toads come from? We could answer that question pragmatically only if we waited for the inevitable drying out of the desert pool and observed what happened to the toads. As it turned out, that was to take two more weeks.

We were amazed to see that innumerable tiny tadpoles were already swimming around in the water, although the pool could not be even a week old. The tadpoles must have developed from eggs in only a few days—an adaptation to the short-lived nature of such desert pools. We later checked this assumption in a scientific paper on the toads and found that it was correct; the eggs of these desert toads develop into tadpoles within two or three days. For the eggs of European toad species, the same process takes at least a week.

The tadpoles grew with incredible rapidity; they apparently fed on algae that had also formed in large quantities in the desert pool. We found these days of study full of drama. We could literally perceive that the creatures were engaged in a life-and-death race with the steadily dropping water level of the pool. Would the tadpoles succeed in completing their metamorphosis into toads before the pool dried out? If not, they were doomed to suffocation, for in this phase of their lives they have only gills with which to extract vital oxygen from the water. Not until they are mature toads do they have lungs with which to breathe air.

As the water continued to evaporate, the tadpoles displayed an interesting type of social behavior. Probably in order to find shelter from the sun's rays, which were no longer being filtered by a considerable depth of water, they pressed so closely together that the portion of each tadpole's body exposed to the sunlight was considerably reduced. Then, when the water level dropped so dangerously low that the tadpoles were barely covered, they showed another type of social behavior. Some groups stopped eating and began vigorously using their tails to dig small hollows in the mud at the bottom. The significance of the behavior was self-evident. The tadpoles were trying to delay the drying out of the pool.

During those dramatic last days, when large parts of the pond were almost completely dried out and the dense swarms of tadpoles were concentrated in the few remaining puddles, we observed an interesting change in two other species of desert toads, *Scaphiopus bombifrons* and *Scaphiopus hammondi*. Several of these individuals developed toothlike jaws and shortly began to eat dead tadpoles that had succumbed in the struggle for life. But the toads also attacked living specimens. This occurrence of cannibalism undoubtedly was due to the closeness of the race with death.

By the fifteenth day, almost all the water in our pool had evaporated. In the evening, the tadpoles that had almost completed their metamorphosis and now had legs lay in swarms pressed close together on the muddy bottom and in the holes they had dug; but there was not enough water to cover them. Were they to die so close to their goal? Once again, the social behavior that had made them swarm in order to shade their bodies saved their lives. By the very density of their concentration, the small creatures kept themselves moist and for a while postponed the fatal drying out. That brief respite, and the little bit of moisture in the small hollows they had dug, sufficed to enable the tiny toads to complete their metamorphosis. By the time the water in the last puddles had evaporated a few hours later, the majority of them had made it; they had become toads and were inhaling oxygen with their lungs. These little toads, with a few withered remnants of their tadpole tails still clinging to them, crawled out of the dried pool and sought shelter under small plants. Here they finished their metamorphosis in a few hours.

We wondered what the next step would be. Because of their water-permeable skins, the toads could not survive as long as a day in the hot, dry desert air. Naturally, the change of surroundings involved a change in their behavior. Within a few hours, these creatures had been transformed from water organisms to desert dwellers. The transition to life on land unquestionably represented the most perilous phase in their lives. Protected by water, the tadpoles had been active during the day; we knew that the adult toads led largely nocturnal lives to escape the sunlight.

On each hind leg of these toads is a small, black, spadelike growth of horn, which has developed out of the warty feet that are characteristic of many species of toads. The horny spade has exactly the same shape as an ordinary metal spade. It is arched, and its lower edge is extremely sharp. This growth is the reason the various American desert toads that belong to the family *Scaphiopus* are called spadefoot toads.

By early the following morning all the spadefoot toads had dug themselves in. We had been fortunate enough to witness a few old, mature toads doing so in the early morning hours and had been able to make a film of the remarkable process.

While it is digging in, the spadefoot toad remains seated in one spot. By shifting the rear half of its body first to one side and then to the other, keeping the hind legs drawn up, it gradually makes a depression in the ground. From the side we had a good view of the way the horn spades of the feet cut into lumps of mud and push the soil away. As the toad dug deeper, it spread its forefeet wide apart and pushed, providing the necessary counterforce. It was the most remarkable movement I have ever observed in an animal. Imagine sitting at a table, spreading your hands as far apart as possible, keeping your trunk and head as immobile as you can manage, and wiggling your backside back and forth on the seat of the chair—that is what the toad's movement was like. Moreover, the operation took only about two and a half minutes.

We had heard from an American scientist friend that no one had ever succeeded in finding a specimen of this type of toad after it had dug itself in. Several times, we were told, the exact spot of the digging had been carefully marked, but a later search of the ground had always proved fruitless. What became of the toads? It was inconceivable that animals dependent on water could survive in the dry ground, especially when years might pass before the next rain.

I hit on the idea of marking a few spadefoot toads, several of which we had kept in captivity for a few days' observation, with a long ribbon that I would tie to a hind leg before they were released. The ribbon was to be as light as possible, smooth, and highly conspicuous—preferably red. We rummaged through our car and searched all our clothing without finding a suitable scrap, so I decided to drive as fast as I could to the nearest small town, about 20 miles away, and look in a store for the right sort of ribbon. Haste was of the essence, since our toads had to dig in this very day, before the soil of the dried pool was baked as hard as stone by the heat.

The saleswoman was dumfounded when I rushed into the small department store in my mud-stained desert togs and blurted out what I needed. A mannequin in the middle of the store helped me out. The plastic model was dressed in a glamorous negligée whose lace collar was held closed by a broad red ribbon tied in a big bow. That was exactly what I had in mind, I told the saleswoman, pointing at the mannequin. Smiling tensely, certain she was dealing with a madman, she told me that the outfit could be bought only as one complete piece. I insisted that all I wanted was the red ribbon, and finally she sold me several yards of bright red silk ribbon. Delighted with my purchase, I ran from the store and drove madly back to the toads in the desert. I can recall the saleswoman staring after me through the shopwindow.

By the time I jumped out of the car at the dried pool, an hour and a half had passed. The morning was still young, however, and there was time enough to set the toads free and let them dig in. But fastening the ribbon to a toad's hing leg was harder than I had imagined. It could not be tied tightly without cutting off the circulation in the toad's leg, but if it was left too loose the toad quickly shook it off. I finally hit on the idea of winding the broad nylon ribbon around the toad's thigh several times before knotting it. That way it held. When we placed the toads on the ground, they began to dig in right where they were. Obviously they were in a great hurry to reach safety from the hot rays of the desert sun. We observed with satisfaction that the ribbon markers were holding. But how deep would the toads dig in? Would frictional resistance become so great as the depth increased that the ribbons would come off? Still, each ribbon was two yards long. Or would the increasing resistance prevent the toads from reaching the depth they wanted? In that case, the experiment would be scientifically useless. We would just have to wait and see. The toads vanished under the surface, and we watched the red ribbons in great suspense as they disappeared bit by bit into the ground. We had tied ribbons to three toads. But after only a few minutes no further movement was observed in any of the three ribbons. Since it was unlikely that all three had come undone at once, the toads must have stopped digging when they were only an inch or so below the surface. There were two possible explanations. Either they did not want to go any deeper or the ribbons were causing them such trouble that they could not dig in any farther.

In order to check the effect of the latter factor, we had carefully marked the spots where a few ribbonless toads had dug themselves in. When the ribbons had remained motionless for a considerable time, we dug one of these other toads out very carefully. It was only an inch or two below the surface. Then we dug out one of our beribboned toads. It, too, was at the same depth. Satisfied that the ribbon seemed to have had no influence, we carefully buried both toads again.

As we could tell by the ribbons, there was no change for the next three days. The toads remained where they were, not far below the surface. We dug up two of them to check. They had an apathetic look, kept their eyes closed, and seemed to be asleep. Apparently they intended to wait for the next rain. But it might be years before an abundant rain fell again. How could the animals possibly survive that long and yet endure the temperature of the heated ground after it was completely dried out? It could get as hot as 70° C. (158° F.), and it would hardly be much cooler so near to the surface. We made plans to come back during the hot dry period a few months later to see if we could find the answer. It would be easy enough to locate the toads, now that they were marked by the red ribbons.

It is here that we return to the beginning of this account: I was digging in the shade of the canvas, hacking open the stone-hard ground in a blazing, light-drenched desert. When we returned to the dried pool three months later, tensely wondering what might have happened to our toads, to our great disappointment we found not one of our red ribbons. What had happened? Had the curious and intelligent coyotes or desert foxes pulled the ribbons out of the ground, delighted with the plump morsels attached to these telltale leads? But we could not even find any traces of the holes in which the toads might have been buried.

Actually, there was only one place for the toads to be. Since we could safely assume that they had not come to the surface again, they could only have dug themselves deeper into the ground. Luckily, we had not entirely relied on our ribbons, but had marked off the area with wooden stakes. We put up our canvas for shade, and then I began digging two probe trenches, each 3 feet deep and about 5 feet long, and meeting at a right angle. That project turned out to be convict's labor. "Digging" was certainly not the word. The deeper my pick drove into the stone-hard ground, the more improbable it seemed that any toad could have survived there. Although we did find slight traces of moisture about a foot below the surface, the ground itself was so hard and dense that any animal would certainly suffocate in it.

When I had combed my two trenches, without finding a trace of a red ribbon, we began scraping down the vertical walls with trowels. If the toads were still there, we were bound to come upon them sooner or later. We had to proceed very carefully, for we did not want to injure any toads that might be in the ground. Such a systematic search was clearly going to involve days of work, but it was the only method that held out any promise of success. We felt more like archaeologists searching for the remains of an ancient civilization than biologists who had taken it into their heads to find some spadefoot toads.

At a depth of about 2 feet my wife came upon a spot that showed signs of previous digging. Very carefully she scraped away the soil and gradually exposed a small hollow space. When the opening was large enough, we were able to see a spadefoot toad in the hollow. The animal looked dead, and even when I held it in my hand we could detect not the faintest sign of life.

Meanwhile, I had exposed another toad in the wall of the second trench. This one showed a feeble pulsing at the throat, at very long intervals, which seemed to indicate that there was slight respiration. When I held the toad in my hand for a while, it opened its eyes. We had found confirmation of what was already known from laboratory experiments. The spadefoot toad is capable of bridging the months and years between rainfalls in a deathlike estivation. The second toad had probably been disturbed and wakened by all our digging.

We placed both toads in a bowl filled with water, and the first specimen also awoke slowly to new life. We were so excited about our find that at first we did not give a thought to the fact that neither of the two toads was marked by a ribbon. Yet we were quite certain that these were our "own" experimental subjects; we had made sure that they dug themselves in at a considerable distance from any others so that there could be no possibility of confusing them. At least one of the two toads we had located must therefore have—or have had—a ribbon. Apparently the ribbons had come loose, after all, as the toads dug deeper and deeper. To reach a depth of 2 feet in that tough clay soil, even while the soil was still wet, the toads must have exerted enormous energy. In the course of their digging they probably had managed to shake off the ribbons. If so, the ends of the ribbons still above the surface could have been sliced off by the sand-laden wind and carried away.

But why, after leaving the pool, had the toads not immediately dug down to their present depth? Why had they lingered for days just below the surface? We thought we had an explanation for that. Since there may be several thunderstorms during the period of summer rains, the toads had paused just under the surface in a waiting position. If another rain had come, they could have reached daylight in a short time. But when no further rain fell and the upper layer of soil continued to dry out, while its temperature steadily rose, the toads acted on that signal to escape from the threatening dry period by digging deeper.

During estivation the metabolism of the toad is so reduced that its vital functions can be detected only with highly sensitive instruments. The consumption of oxygen is so low that the small amounts that diffuse through the dense soil are sufficient to sustain life in these animals.

Extensive studies of spadefoot toads have shown that they can store large quantities of fat in their tissues and large quantities of water in their bladders, in the form of diluted urine. The water stored can amount to 30 per cent of their body weight.

It is interesting that for estivation the toads form a special blackish skin that covers the entire body. Some species of desert insects have specially constructed skins that are permeable to water in one direction. That permits the insects to take in water, but it prevents the precious fluid from evaporating through the skin. The toad's estivation skin probably functions in a similar manner, protecting the animal from drying out but allowing water to enter. Such a characteristic is essential for bringing estivation to an end. When sufficient rain falls, after one or several years, water soaks down to the toads deep in the ground. They promptly wake up and dig their way through the softened soil to the surface.

It is characteristic of relief deserts that precipitation is extremely uneven, in both time and amount. Each of the many species of spadefoot toads has adapted to the special conditions of its distribution area. The rainfall in a given area penetrates the ground to a specific depth. The water's maximum penetration depth during a region's principal rainy period corresponds to the depth to which the spadefoot toad population of that part of the desert digs in. Only the heaviest rainfalls reach the toads and stimulate them to dig their way to the surface.

Animals are not the only organisms that fall into a state of dormancy lasting for several years. A comparable phenomenon can be observed in desert plants. When there is no rainfall for years at a time, plants cannot carry on photosynthesis, the most important phase of their metabolic processes, because the requisite raw material, water, is not available. The parts of the plant aboveground wither and die, but the plant's life processes continue underground. In specially developed tubers and bulbs, the plant stores up food and water reserves that permit it to survive for years. In principle, these desert plants do what many of our garden plants—tulips, for instance—do when they temporarily retreat from life aboveground for the winter. The underground food and water reservoir of some desert plants is often of great size, permitting it to survive many years of drought. The storage chamber of the tsama melon (*Citrullus*), which grows in the Kalahari Desert of South Africa, reaches the size of a football. The Bushmen, who call the plant *bi*, carefully note the location of the vines. When all their water holes have dried out during a long period of drought, they dig up the big roots of the plants.

The small madar tree (*Calotropis*) of the Sahara, which reaches a height of 10 feet, is capable of going into a dormant phase that will see it through droughts of six to eight years. The tree does not let its aboveground parts die during the dry years, since it could not regrow its thick trunk and branches after every period of rainfall. However, the tree has developed a watertight, corklike bark, which, besides, is almost white, so that it reflects sunlight. Moreover, the tree shades itself during the years of drought by using its own leaves for a parasol. The madar has unusually large and thick leaves for a desert plant—they reach a diameter of 8 inches—and it does not drop them during dry spells. The leaves, a juicy green after rain, do die completely during the dry period, and look yellow and withered, but they remain clinging to the branches (see Plate 17). Since the leaves have a tough, leathery quality, they persist for years. Clinging as they do, they shade the trunk and branches from the vertical rays of the sun, as well as the ground at the foot of the tree, protecting the roots in which food and water are stored. When, after a period of years, there is fresh rain, the tree grows new leaves; only then do some of the old leaves drop.

Poor-wills and jerboas, which hibernate or estivate once or several times in the course of a year, have at least one active phase of life per year during which they reproduce. The spadefoot toads, on the other hand, and such plants as the madar are incapable of reproducing every year. A hostile environment has forced them to suspend their functions for years at a time, and there is an interruption in the normal succession of generations, ordinarily a prerequisite for the preservation of the species.

The fact that the plants and the spadefoot toads still exist is evidence that these life forms have so far been able to weather the crisis to which their species have been exposed. A permanent break in the succession of generations, which would mean the end of the species, has been postponed for a while.

If we view the behavior of the poor-wills and the jerboas as a first step in life's retreat from the desert, the inability of some life forms of the desert to reproduce every year represents the second step in that retreat.

What will happen to life forms in the future if, in the course of the earth's development, the continents become more and more arid, more and more hostile to life; if the intervals between rainfalls stretch out longer and longer? We need not wait several million years to witness this drama; it is already in full swing in the desert. We owe our opportunity to look into the future to the fact that not all living organisms are at the same stage of evolution, nor do they react in the same way to changes in their environment. If that were not so, there would be only a single life form left on earth, the one that was the furthest evolved. That probably would be man, since in geohistorical terms he represents the most highly evolved species. But the nature of human beings plainly shows that no life form can exist alone. Man is dependent on plants not only for food but also for the indispensable oxygen he breathes. He is even dependent on bacteria in his intestines to help him digest his food.

Finally, man is dependent on the rest of nature in quite a different sense—one that at first glance seems to lie outside his vital "practical" requirements. It is not incidental that we derive aesthetic pleasure from the varied forms of life that share our environment. In fact, the presence of living nature appears to be essential to our psychological well-being. All art and myth testify to this primeval human impulse. Without the sight of a bird flying, a gazelle racing over the plains, a blossoming flower, we would languish. Spiritually as well as physically, man is not so constituted that he could survive alone on a barren planet spinning through an empty universe.

The life forms existing today run virtually the full gamut of the animate history of the earth. Examples of earlier phases of evolution still exist as species, as living fossils. The tuatara, a lizardlike reptile that survives on outlying islands off the coast of New Zealand, dates back 150 million years to the time of the dinosaurs, when tropical jungles overspread the whole world. The lungfish recalls that period 350 million years ago when the first vertebrates conquered the land. And the most numerous of all the life forms on earth, the algae and bacteria, have been around for more than 3 billion years. They stand at the beginning of all life.

The desert regions of the earth present many gradations of hardship, from the dry steppes, which receive regular rainfall but at long intervals, to the very heart of the desert, where total aridity reigns. The manifold forms of life to be found there—plants, insects, amphibians, birds, and mammals—have varying periods of evolutionary history behind them and react in sundry ways to their challenging environment. The sand grouse seem to have adapted perfectly; the spadefoot toads have started to retreat. We are observing the possible course life may take millions of years from now as all the continents revert to desert.

## Life on Call

Bill Robertson, a ranger, was driving his jeep through a remote part of the Mojave Desert. For twenty years he had been making a regular monthly patrol that carried him to some of the least-known corners of that region. Normally he made his tour in the middle of the month. But this time he had been delayed because the jeep needed motor repairs, and it was now the end of May. He was unpleasantly aware that he had set out ten days later than usual. At this season the temperature rose from day to day. He noted in his logbook that it was 40° C. (104° F.) in the shade. One of his duties was to check the automatic weather stations.

The part of the Mojave Desert he was driving through has one of the fiercest climates on this planet. A few years ago the highest air temperature in the shade ever recorded on earth was measured here: 52° C. (123° F.). Robertson was driving over a gigantic plain completely surrounded by high mountain ranges. The ground consisted of sand or of great stretches of black rock cracking from the effects of heat and dryness. Among the sand and scree lay extensive basins, their clayey soil a network of fissures caused by drying and shrinking. It was a veritable landscape of hell. On this particular day, the high black mountains surrounding the valley could not even be seen. Clouds of sand and dust, whirled up by sharp gusts of wind, limited visibility. Frequently Robertson could see no farther than a few yards. Then, at a moment when the air cleared briefly, he observed a remarkable phenomenon. Never had he seen the like. A short distance above the desert floor floated a curious image in the form of a long, flat lens of a blue color. The movements of the quivering hot air seemed to have been communicated to this image. The ranger shifted into four-wheel drive and drove slowly toward the strange sight. The hot, shimmering air made it impossible to estimate how far away it was. Also, the phenomenon disappeared for minutes at a time behind clouds of whirling sand and dust. A few miles closer, the lower part of the lens appeared to be in contact with the ground, instead of just above it. Then, as he approached that point of contact, which at first had looked like the stem of a mushroom, the lens widened and widened, and the blue color seemed to be flowing out of a vast floating bubble in one of the hollows of the ground. When Robertson stopped, he found himself on the shore of a blue lake.

In this dip, which was protected from the wind, evaporation had produced a cool layer of air directly above the water, and over this cool stratum the wind had driven hot air. The space between the two layers of air, or, rather, the underside of the hot stratum, was acting like a mirror and reflecting the water of the lake. What Robertson had seen was an optical illusion. It was the exact opposite of what happens in a conventional mirage. The latter, a well-known and common phenomenon, occurs when sunlight strikes the top of a thin, hot stratum of air that has formed just above the ground. The sunlight is reflected back into a cooler layer of air above the hot layer. The desert traveler who imagines he sees a blue lake is, in fact, seeing only the reflection of the blue sky. The swelling movements of the hot air deceive him into seeing ripples and waves. Any rocks that stick out through the reflective layer of hot air seem to be floating in the air and often assume fantastically distorted forms. That would account for some of the elaborate visions reported by desert travelers—visions of houses and fabulous palaces, even whole cities, on the shores of azure lakes.

When Bill Robertson stepped to the edge of the lake, again he could not believe his eyes. The water was swarming with millions of tiny shrimps. The lake had formed only a few days ago, after a brief but violent local rainstorm—probably the first rainfall in years. The last recorded rain had been twenty-six years ago.

Rain water had collected in the basin. And there, now that the prime requisite for all life was present, millions of shrimps filled the water. How such a "miracle" could have taken place was not fully explained until years afterward.

The ranger filled a canister with a sample of water and several shrimps and continued on his patrol. The discovery had been sheer chance. If his jeep had not needed repairs, Robertson would have set out on his patrol in the middle of the month and would never have discovered the lake and its crustaceans. When he visited the basin on his next round, all the water had evaporated and the bottom of the short-lived lake was a maze of deep cracks. There was no sign whatever of the shrimps. The wind probably had carried away the dead, parched bodies of the little creatures.

Consider the incident from the standpoint of the mathematics of probability. For twenty years the ranger had been making his monthly tours—a total of 240 of them. Assuming that he stayed in the service and continued his patrols for another ten years, that would make an additional 120 tours. Judging by the record of his first twenty years, it was unlikely that he would ever again discover a lake with shrimps in it. Overall, he would have made some 360 tours in that part of the desert and only once have encountered a body of water filled with shrimps—and even then by sheer chance, simply because his jeep had needed repairs.

In the spring of 1973 I had an opportunity to visit the area in which Bill Robertson had discovered those miraculous shrimps in the desert lake. The basin was riddled with cracks from the long-ago parching of the wet mud. The sun had baked the ground to the hardness of stone, and there was no indication that any water had stood here for years.

I had conceived a simple experiment that I hoped would recreate the "miracle." Since I could not, of course, conjure up a rainfall to fill the basin with water, I first hacked a few rocklike fragments of earth out of the hard lake floor with a pickax. I put these lumps of earth into a tall glass and then provided an artificial rainfall by filling the glass with pure water from my canteen.

Two days later, I held the glass containing lumps of earth, which had meanwhile softened to mud, against the light, and I saw dozens of tiny desert crustaceans swimming around in the water. The shrimps, only about a millimeter long, could just be made out with the naked eye. In only two days, these creatures, which in shape resemble tadpoles (for which reason they are called tadpole shrimps), apparently had developed out of nothing, out of mere mud. I had sifted my lumps of earth beforehand to check whether there were any specimens in them already, possibly estivating, like the spadefoot toads, and ready to be awakened in the presence of water. Although I had examined all the samples of earth under a magnifying glass, I had discovered nothing but particles of mud and grains of sand—nothing that bore any resemblance to a minute shrimp. Also, I knew that full-grown tadpole shrimps reach a length of about 3 centimeters and therefore could not be missed. There was no question but that the creatures in my glass were newly developed young.

To continue the experiment, I placed the glass in the sun and waited for my artificial lake to evaporate. As the water level fell by about two centimeters a day, the shrimps grew at a truly incredible rate. They fed on algae and microörganisms that were also developing in the glass. After a week, by which time the column of water was half evaporated, the shrimps were a centimeter and a half long. The space available for them grew more and more cramped, and every day some of them died, since their increasing appetites could not be satisfied. The surviving animals grew all the faster as the competition diminished. The water level sank inexorably, and it was a matter of only a few days before these desert crustaceans were doomed. What would become of them? It was obvious that the shrimps did not save themselves by going into estivation; otherwise, I would have found some adults in the dry soil. So I concluded that any descendants of my nearly adult shrimps now swimming around in the glass would also be unable to estivate. Yet the species somehow preserved itself over many years of drought; Bill Robertson's observations and my own experiment had proved that. The shrimps swimming around in my glass could not have sprung by spontaneous generation out of sand and water molecules.

By the twelfth day, the shrimps were three centimeters long, and hundreds of tiny eggs had formed on their undersides. By noon of the fourteenth day practically all the water had evaporated. The six specimens that still survived now displayed an interesting form of behavior, like the kind I had observed during the metamorphosis of the young spadefoot toads. The shrimps thronged close together and, since they were no longer covered by water, began vigorously twisting their bodies to make a depression in the mud. The little remaining water collected in this hollow. The shrimps used the last moment of their lives to lay their eggs in the mud. The determination to preserve their own lives as long as possible had not just gained them a few minutes; it had enabled them to lay their eggs in a protective depression a few centimeters below the surface of my artificial pool, which had now become desert again. In so doing, the survivors were carrying on their species in behalf of all those that had not been able to complete their life cycles.

Desert shrimps are unable to survive for prolonged periods in the desert or to overleap years of drought by such an adjustment as temporary dormancy. They often exist for only one or two weeks out of a period of many years or decades. But their eggs are able to survive for decades. When rain comes again, the next generation of shrimps hatches from the eggs. These eggs that possess such amazing viability have the size and appearance of grains of sand. No wonder they had escaped my notice when I examined the soil samples under a magnifying glass. Those eggs had produced the "miracle" that Bill Robertson witnessed and that I observed in my water glass.

A good two weeks after beginning my experiment, I placed my samples back on the floor of the basin from which I had hacked them out. They were just a few dry, stony lumps of earth, in which the new eggs could not be differentiated from grains of sand. The dried bodies of the dead shrimps had been carried off by the wind.

The rainfalls that make the brief lives of these creatures possible often occur as the result of favorable general weather conditions. When such conditions prevail, several brief rains may fall in rapid succession, and the lakes that are formed will last long enough to serve for perhaps two generations of shrimps. During such times, which come perhaps twice in a century, the species can multiply considerably, so that by the end of the rainy phase the floor of the dry basin has become a vast repository of eggs for future generations of shrimps.

If rain does not come for a very long period of time, the viability limit of the eggs may be exceeded, and that could lead to the extinction of the shrimp population. To avert such a catastrophe, the species has a last resort. Amazing as it sounds, the eggs themselves sooner or later go hunting for water, often to a remote region of the desert where a downpour has filled some other basin with water. Over many years the desert wind wears away the dry lake bottom, millimeter by millimeter, and carries it away. When, someday, the layer containing the eggs is exposed, the tiny grains of life are blown out of their basin and carried great distances by the wind. The eggs are superbly fitted for such a journey; they are encased in the toughest of shells. Even if only a hundred out of the millions of eggs carried by the wind reach a pool or a crack in the rock that holds water, a new generation will develop.

If all the eggs were blown out of the original basin and none of them reached another water hole, the species would not survive. But insurance against that is provided by the many deep cracks that form in the lake bottom as the ground dries out. The same wind that carries the eggs away blows thousands of others into these fissures, so

that a last reserve supply remains in the place where the eggs originated.

Some American scientists think that the eggs of the tadpole shrimp
may remain viable for as long as a century. But future generations of
scientists will have to confirm that supposition, when, a hundred
years from now, one of them drops into water the eggs that have been
put into storage for that purpose.

Some of the most striking, most beautiful, and also most numerous
forms of life in the desert are rarely seen. I had an encounter with
one of those remarkable forms of life in the spring of 1973 near the
Mexican border. The heaviest winter rains recorded in the region
since 1915 had occurred shortly before, and a friend who was an
American desert scientist had advised me that I was about to witness
a spectacle such as I had never seen. A few weeks later, I knew he
had not exaggerated. The desert I was seeing was no longer a desert.
The stony, sun-parched ground that extended between giant cacti
and mesquite trees was completely covered with a carpet of flowers.
The blue blossoms of the desert lupine and the golden yellow and
orange calyxes of the goldpoppy, or desert poppy, were crowded
together so closely that it was impossible to walk through the patch
of desert without tramping on dozens of flowers at every step. The
carpet of blossoms, which stretched on to the horizon, consisted of
ephemeral plants, as they are technically called. These are annuals
that cannot survive permanently in the desert, either by adaptation or
by temporarily reducing their metabolic functions. They cannot withstand the years of drought that follow the occasional rains. As soon
as the seedlings raise their heads above the desert floor, a life-and-
death race begins between them and the withering sunlight. The
plants can live only a few days, while the ground is still moist; they
must flower almost immediately or not at all. The sea of blossoms
lures flying insects in great numbers. They move into the desert, often
from remote areas, to take advantage of the unique supply of food
from billions of nectar-filled flowers. The clouds of insects hovering
above the blossoms quickly pollinate the short-lived beauties. By the
time their few days are over and their blossoms withered, the plants
have formed seeds, which represent their sole chance for survival.
Like the eggs of the desert shrimps, the seeds remain viable for years,
even decades, embedded in the desert soil. After a rainfall the next
generation of plants springs to life. A few months later, in the sum-

mer, when I revisited the part of the desert where I had seen that miraculous flowering, I found only a few wilted remains of the desert annuals. But I could not see the billions of tiny seeds hidden among the stones and grains of sand waiting for the next rain (see Plate 34).

For such life to function properly, the seeds of the annuals must germinate only when there is enough rain, and enough soil moisture, to enable them to complete their growth. If a small shower were to cause the seeds to germinate, they would be wasted and the species would become extinct.

It is essentially the same problem as that faced by the spadefoot toads, whose reproductive cycle is likewise dependent on a sufficient amount of water. In order to avoid being tricked by an inadequate rainfall, every spadefoot toad population has a certain depth to which it digs itself in, a depth that corresponds exactly to the penetration depth of the heaviest rainfalls in its distribution area. But plant seeds cannot take such precautions.

The seeds lie right on the surface of the desert, among small pebbles, or, at best, are covered by a thin layer of sand, and they are moistened by the lightest shower. But they do not germinate. Amazingly enough, even when there has been sufficient rain many of the numerous species of ephemeral plants do not germinate. Great quantities of seeds lie hidden in the desert soil, and if all of them germinated at once there would not be room for the plants to grow. Then how does the moistened seed of a goldpoppy know when sufficient rain has fallen for it to reach its full growth, and how does it learn how its competitors are reacting—whether, for example, desert daisies or purplemat are using the rain for their own rapid growth, and the goldpoppy seed would be wiser to wait for another rainy period before germinating?

In parts of California's Mojave Desert and the adjacent Sonora Desert of Arizona, attempts have been made to find answers to those questions by experiments outdoors and in greenhouses. These regions may have precipitation twice in a year. It was discovered that the seeds of certain annuals germinate only in the spring, after the winter rainfalls. Other species sprout only in the fall. Also, many experiments have shown that the seeds of most of these annuals do not germinate unless they have been supplied with water equivalent to a natural rainfall of about an inch. The seeds lying on the surface of the desert possess a growth-inhibiting factor in the seed coat, which is dissolved only by that amount of water. In this way the seeds protect themselves from germinating after brief showers that do not provide enough water for them to grow and flower.

In addition, greenhouse experiments have shown that there is a close connection between the amount of water and the ground temperature. The seeds of the species that react to winter rainfalls germinate only when the ground temperature is low, as it is under natural conditions in the spring. When the ground temperature is raised to the summer level and the same amount of water is supplied, the seeds do not germinate. The experiments satisfactorily explain why one group of ephemeral plants sprouts in the spring, another in the fall. Only the proper combination of temperature and precipitation provides the stimulus, different for each species, that initiates germination and growth. The ephemeral plants have developed their own modes of preventing overpopulation when sufficient rainfall does occur in their normally hostile environment. Once again, life has succeeded in making the best use of the space available to it and of the opportunities and limits of an arid environment.

But even when precipitation and ground temperatures are right and provide the best possible conditions of growth for a given species, some of the seeds of that species will remain in the soil without germinating. It is as if these seeds did not trust the weather. They constitute a reserve, and a necessary one. For, in spite of initially favorable growth conditions, a sudden heat wave can always descend on the desert, and the plants that have already germinated will die before they have had a chance to bloom and mature the next generation of seeds. If nature had not wisely provided a reserve of seeds, the entire population of a given plant would be condemned to extinction. Even if the same kind of disaster happens in the next period of favorable growth conditions, a further reserve is left—seeds that have a still more pronounced habit of strategically delayed growth and will preserve the species for many more years.

Some dry and particularly hostile regions of the desert have ephemeral plant populations that blossom only a few times in a century. The generational delay in their case, as in that of the desert shrimps, can amount to decades. Such a long interval in the succession of generations constitutes a serious danger to the preservation of the species. For the seeds of the desert plants, like the eggs of the desert shrimps, cannot keep indefinitely; at some point they will lose their viability. In the interior of the Sahara, for example, when rain comes after perhaps thirty years of dryness, even the heaviest downpour rouses not a single ephemeral plant to life. And if, someday in the distant future, rainfall should cease on the continents for a century or more, the fate of the last highly evolved species of plants would have been sealed.

The most dramatic consequence of the retreat of plants from the increasingly arid continents will be the lessening of photosynthesis, and of its oxygen production. Certain plants of the desert already exhibit the first step in that dramatic evolution—for example, the metamorphosis of trees into cacti, which has led to the atrophy of the leafy areas in which photosynthesis is carried on. The second step in this development is exemplified by the dormancy of some desert plants, which for years at a time do not engage in photosynthesis. The third step is represented by the long pauses between generations of the ephemeral plants; they often exist as oxygen-producing organisms for only a few weeks at intervals of decades. But, as we shall see, the scarcity of oxygen-producing plants in the steadily spreading deserts is not the worst aspect of the danger. What is far more serious is that the deserts themselves have become the greatest consumers of oxygen on our planet.

## Microforests of the Desert

I am about to deal with the last—and in many respects the most portentous—step in the retreat of life forms from the desert, and I shall begin with a riddle. As it happens, the riddle represents a terse but precise description of one of the most interesting symbiotic life forms of the desert.

It is about the size of a pullet egg but is angular in shape, and is black on the outside and light-colored on the inside, sometimes almost white. It can be so hot that you burn your hand if you try to pick it up. But sometimes you do not burn your hand, and you see that it has six legs and two eyes.

Since I think it unlikely that most of my readers will guess the answer to this riddle—although a few may—I intend to provide the solution. But to do so requires an elaborate description of this symbiotic life form.

All deserts—indeed, all the arid regions of our earth—are distinguished by a reddish to black surface coloration. Any desert traveler will remember the vast expanses chiefly through their coloration—the reddish yellow seas of dunes and the gloomy, blackish brown rocky wastes, the hammadas. Photographs of our planet taken from space show that, aside from relatively small green areas of vegetation, by far the greater part of the continental surface has the same coloration. The reddish color of the arid areas is suggestive of scorched earth and burned rocks. This association we make with the color red comes partly because we know from experience that the red flame of a fire is hot. And the red flame merely shows us what is taking place during oxidation, or combustion: the combining of oxygen with other elements.

As we already know, oxygen is extremely active chemically. Almost any material, even metals, will burn in pure oxygen. Much of the barren desert surface is covered with stable compounds of oxygen, which give the desert its coloration. These oxides are compounds of atmospheric oxygen with such metals as iron and manganese, enormous quantities of which are contained in the earth's crust. While most materials, such as wood or coal, must be heated to a minimum temperature, called the ignition point, in order for combustion to take place, some materials, including metals, will oxidize at ordinary temperatures. For iron and manganese oxides to be produced, the metals have only to come into contact with atmospheric oxygen.

A particularly impressive example of a desert surface consisting of metallic oxides is to be found in the southern part of Morocco, in the northwestern Sahara. That is the region in which my wife and I carried on our bird studies for a number of years. Here is to be found an extensive flat hammada with a surface of small pebbles that is black in color, in contrast to the largely reddish rocky wastes of most of the Sahara. The black coloration comes from the high percentage of manganese oxides mixed in with iron oxides. The black landscape of rock is bounded on the east by enormous reddish yellow sand dunes. The conjunction of the two colors, and the vastness of the area covered by each, make the region one of the most striking in the Sahara (see Plate 36).

Because of the black coloration, the ground temperature during the summer months is the highest known anywhere on earth as the result of solar radiation alone. I have measured more than 80° C. (176° F.) in the summer. The stones are so hot they burn the fingers.

In that level stony surface the eye cannot discern a plant, a creature of any kind. Water spilled from a bottle onto the rocks evaporates as if it had struck the hot lid of a stove. Then angular stones, an inch to an inch and a half long, which form a flat mosaic pavement, have highly polished surfaces. They look as if they had been coated with black or russet lacquer. When the sun falls at a slant on this surface, each of the polished facets reflects the light. All the way to the horizon, the desert seems to consist of billions of flashing silver slivers (see Plate 35).

The notion that the stones are covered with a layer of lacquer is appropriate, for when a stone is removed from the black mosaic a bright gap appears in the pattern of the desert floor. Only the stones on the surface are dark-colored; the rock underneath is light.

We made frequent use of that feature in our ornithological studies. We just removed the top black stones and we had markings on the desert floor—arrows showing flight direction, for instance—that could be seen from a great distance. And the idea of a coating is reinforced when the black stones are smashed with a hammer. The broken faces show that the stones are black only on the outside; inside they are a light color, often almost white.

The black or reddish brown metallic oxides enclose the stones like a peel. Here is part of the solution to our riddle. That leaves only the legs and eyes to find out about.

But how were the metallic oxide crusts formed? Until a few years ago, desertologists, if we may invent a useful word, thought that the inanimate forces of weathering alone were responsible for crustal formations of this kind. They believed that tiny amounts of moisture, either from the rare rainfalls or from dew, penetrated hairline cracks in the rock, crevices that already existed as a result of the extreme temperature difference between day and night. The moisture dissolved some of the metallic oxides present inside the stone. Then the moisture was drawn back to the surface by the sun's heat, and when it evaporated the iron and manganese oxides were left behind as a thin film on the outside of the stone. In the course of many years, a crust formed, eventually becoming an inch or so thick. Exposed on the surface to the night dew, the manganese and iron oxides absorbed water and were transformed into hydroxides. Or, more simply put, they rusted. If that analysis was correct, the metallic glitter, the desert lacquer, did not arise from the polishing of wind and sand, as used to be thought, but from the layered fine structure of the crusts themselves.

Over very long periods of time the inside of the rocks or stones would be more and more broken down by this process, while the outer crust thickened. Temperature variations would tend to be extreme, because the outside of the stones or rocks, being dark, absorbed a maximum of heat. Someday the difference in temperature would cause the crust to burst. Then the sand-laden desert wind would penetrate the cracks and splits quite easily and would sandblast the softened interior of the stone. Eventually nothing would be left but the hard, resistant shell of metallic hydroxides.

It is quite common in a hammada to find stones the size of footballs that are hollow inside. The shell plainly shows the broad cracks through which the interior was sandblasted. When these hollow balls of rock are thrown to the ground, they smash with a sound like that of a breaking bottle. The surface of the hammada is often covered with metallic hydroxide shards from hollow stones that have been shattered by the natural forces of weathering. Driving a vehicle over such terrain, you have the feeling that you are riding over broken glass.

Then, a few years ago, a sensational discovery was made by an Israeli institute for desert research. The Israelis had built mobile laboratories that were equipped with such sophisticated instruments as infrared spectrographs and devices for measuring light, temperature, and moisture conditions at desert sites. As one of their projects, the Israelis subjected the stones coated with desert lacquer to chemical analysis.

Immediately after the night dew dampened the stones with tiny amounts of moisture, the instruments began to register something exciting. The stones were beginning to breathe; they came to life. They inhaled oxygen and exhaled carbon dioxide! At sunrise the stones reversed this vital process; they took in carbon dioxide from the air. That could only mean that they were practicing photosynthesis, were, like green plants, using sunlight to produce food. But during the early morning, as the temperature rose and the last traces of nocturnal dew evaporated, the photosynthesis stopped and the stones began breathing oxygen again.

The conclusion was obvious and unequivocal. Those desert stones coated with metallic hydroxides must be plants.

Studied with a strong magnifying glass or under a microscope, the stones of the black hammada do indeed reveal a surface overgrown with plant life! The plants are algae, lichens, bacteria, and fungi; in essence, the whole living complex constitutes a microforest invisible to the naked eye. This microforest has learned how to satisfy its need for water from the minimal moisture supplied by the rare rains and the more frequent dew. Perhaps we should not be astonished that microflora can grow on a dry surface consisting of iron and manganese hydroxides; microörganisms have proved enormously tough in other areas. They live both in the boiling geysers of volcanically active regions and under the Antarctic ice. Several tiny fungi and bacteria have chosen the fuel tanks of jet planes for their biosphere. Microörganisms attached to the outside of rockets have survived unharmed the deadly radiation and the vacuum of a journey through space. Some bacteria in Australia seem recently to have developed an interest in asphalt roads. They are the petroleum bacteria, which can convert asphalt made of petroleum into protein. The bacteria's appetite is so great that they eat their own weight in asphalt every five seconds. Since each of them can produce 17 billion descendants every twenty-four hours, they represent a serious danger to Australian roads.

Microörganisms came into being at the beginning of the earth's story, and they will surely be there at the end. In the distant future, they will be the last remaining life forms on earth. If we may generalize, the simpler and more primitive a living organism is, the more resistant it is to changes in its environment and the greater its chances of survival.

Seen in that light, it is hardly a surprise that microörganisms have found favorable conditions in a desert where more highly developed forms of life cannot exist.

One day, while we were studying the stones in our research area for microflora, my wife called out to me that one of the stones at her feet had moved suddenly; in fact, it had taken a fairly big leap. Our years in the desert had convinced us that nothing was impossible— this was after I had seen the moving rocks of the Mojave Desert— and I quickly ran over to where she stood. She was adamant about her unlikely observation. But we could discover nothing, although both of us, on our knees, searched every square inch of the surface where she had been standing. Each small black angular stone looked like every other. Besides, all of them fitted exactly into the mosaic of the pavement, and for this reason alone they could not have manifested a life of their own. Suddenly my wife saw another pebble stir, and this time she managed to keep her eye on it. When we examined it closely, we saw that it had six legs, two eyes, and a pair of shriveled wings (see Plate 37). It was not a pebble but a locust, which we then and there named the manganese locust. The insect perfectly resembled the stones and pebbles among which it lived. Its black body likewise seemed coated with the desert's lacquer of manganese hydroxide, and it had the same angular shape as the black pebbles. The camouflage went so far that the insect imitated the fine surface structure of the stones; even its large eyes were adapted to the coloration of the environment with their fine black dots. Here was protective coloration carried to an extreme. Locusts are noted the world over for their ability to adapt to their surroundings. There are species that can barely be distinguished from a thin, dried twig. Other species assume the shape, color, and veining of a leaf. Still others look like the blossom of a tropical orchid. Protective coloration plus the ability to reproduce the shape of an object in the environment is known in zoology as mimesis, and it is believed to result from mutations acting in conjunction with natural selection.

In the case of our manganese locust, any selection pressure exerted by enemies seemed inconceivable. What other species of animal could exist in this terrible desert? For that matter, how did the locusts themselves exist here, enduring ground temperatures of over 80° C. (176° F.)? Aside from the temperature of the rock, their black coloration ought to heat their bodies in sunlight to a temperature absolutely fatal to any complex organism. And what did the insects feed on? All those questions would take a long time to answer. I already knew, from my studies of the sand grouse, what diverse means desert animals employ to cool their bodies. I hope that someday I shall have an opportunity to study the lives of these exceedingly interesting desert insects. On the basis of my preliminary observations, I would guess that the manganese locust spends the hot summer underground in a larval state, protected from the sun and heat. I would also imagine that the manganese locust feeds on the microflora on the stones.

But let us return to the question of the origin of the metallic hydroxide crust on the stones. Is it possible that there is a connection between the crusts and the lichens, algae, and fungi that settle on them? The German geologist Wolfgang Wunderlich, who studied these microflora in the Negev desert of Israel a few years ago, made the important discovery that the microflora themselves contribute substantially toward the formation of the metallic hydroxide crust. He demonstrated that the various species of lichens are responsible for transporting the metallic oxides from the interior of the stones to the surface.

Lichens are primitive plants that are distributed throughout the world. They have proved extremely hardy and resistant to all sorts of climates. Some large species form extensive crusts on the surface of rocks. Wherever poor living conditions prevent the growth of more highly evolved kinds of plants, lichens represent the principal type of plant life. They grow on the ridges of the highest mountains, in the icy wastes of arctic regions, and in the hottest deserts.

If we study one of these remarkable plants under a microscope, we can see that the lichen's seemingly unitary vegetable body is not one plant, but consists of two entirely different species of plant that have united for their mutual benefit. They are an alga and a fungus, whose threadlike hyphae surround the alga like a net or, in some species, penetrate into it.

The algae are green plants and produce their food by photosynthesis. The fungi consist essentially of a web of tiny threads; these are the hyphae, which can be seen with the naked eye when, for example, they appear in the form of mold on old bread. As opposed to green plants, the fungi have no chlorophyll and therefore are unable to use photosynthesis to manufacture their food. Like bacteria, which also belong to the vegetable kingdom, they feed by breaking down organic substances, such as the dead parts of other plants in the soil.

It was highly practical of nature to have created a group of plants like the lichens, in which the fungi live in symbiosis with the photosynthesizing algae. The algae supply the organic nutriments; the fungi provide water and mineral nutriments. This mutualism, as it is called, confers on both partners abilities they would not possess alone —for example, the ability to settle on inhospitable cold or hot rocks.

Lichens growing on desert rocks actively draw their nutriments

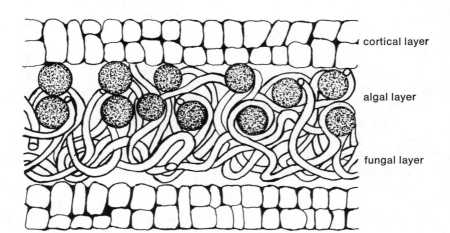

cortical layer

algal layer

fungal layer

Cross-section of a lichen

from the minerals in the stone. The hyphae of the fungus can penetrate into microscopically small crevices and, by excreting acids, dissolve the vital mineral salts in the rock. In utilizing energy to extract minerals, lichens are far ahead of other plants. They are even able to overcome the binding energy in the crystal lattices of minerals. When the hyphae have penetrated deeply into the rock, bacteria, already present in large numbers, produce additional acid, which speeds the dissolving of the minerals. Along with the nutriments drawn from the rock, the metallic oxides reach the surface of the stone, where they are deposited in the form of a crust and change to hydroxides.

The inanimate forces of weathering, such as temperature cracking, the penetration of moisture, and evaporation, do indeed build up metallic hydroxides on the surface of desert rocks. But the microflora make an essential contribution to the process. With that in mind, we come to the crucial point: the problem of maintaining free, unbound oxygen in the atmosphere—without which no highly developed life on the continents would be possible.

Let us recall some of what we discussed in chapter 2 about the origin of the atmosphere. Modern scientists are fairly certain that all the oxygen released throughout the earth's history and accumulated in the atmosphere is a by-product of photosynthesis. Since the carbohydrates produced by plants contain less oxygen than their initial components of water and carbon dioxide, some oxygen is left over as a waste product of photosynthesis. That oxygen can accumulate in the atmosphere if the organic materials containing carbon are not completely reoxidized. That is, oxygen remains only if a proportionate amount of carbon is bound up in the sediments of the earth's crust.

To calculate the total amount of atmospheric oxygen produced throughout time by the process of photosynthesis, we may assume that for every *atom* of carbon buried in the earth's crust and not reoxidized, one oxygen *molecule* was released, or, rather, left over.

Although the oxygen released by photosynthesis actually comes from the splitting of water molecules, the proportion of carbon to oxygen must correspond to the proportion of the two elements in the carbon dioxide molecule. This means that the total amount of carbon stored in the earth's crust today should be matched by a proportionate amount of free oxygen.

Geological studies have determined the amount of carbon with fair accuracy. The sediments of the crust contain, on the average, 0.5 per cent of organic carbon. For the whole earth, this equals an inconceivable $10 \times 10^{21}$ grams. Highly concentrated carbon in the form of coal or petroleum constitutes only a tiny fraction of that amount. The greater part of it is thinly distributed in sedimentary rocks. For the total amount of carbon, there ought to be a corresponding amount of free oxygen, in the atmosphere and dissolved in the water of lakes, rivers, and oceans. And the proportion of the two elements should be the same as their proportion in the carbon dioxide molecule, $CO_2$. In terms of the number of atoms, the proportion would be 1:2; in terms of atomic weight, it would be 12:32. That is, since there is more oxygen than carbon in the carbon dioxide molecule, there should be more free oxygen present than there is carbon bound up in sediments. The amount of free oxygen ought to be about $27 \times 10^{21}$ grams.

But calculations have shown that no such enormous quantity of oxygen is present. Only $1.3 \times 10^{21}$ grams of free oxygen have been found. What accounts for the vast discrepancy?

We know about the readiness of oxygen to combine with other elements. Indeed, recent calculations have shown that the missing free oxygen is bound up in the earth's crust in the form of ferric oxides and sulfates. Only 5 per cent of the oxygen produced by photosynthesis is left today in the atmospheric reservoir of free oxygen. By far the greater part of it has already been bound up in the secondary reservoir of the earth's crust, and there is plenty of room for the remaining 5 per cent to be stored there. The crust has enormous reserves of oxidizable materials, which could consume considerably more oxygen than is available. The magmatic rocks that have reached the surface and been exposed contain a vast supply of bivalent (oxidizable) iron waiting to absorb oxygen. In addition, about half the sulfur distributed throughout the crust is still in the form of sulfides and by consuming oxygen could be transformed into sulfates.

Suppose that the entire present production of oxygen on earth could be stopped by means of some vast and fantastic experiment. The free atmospheric oxygen would be consumed by the crust in a remarkably short time, perhaps only a few millennia. The desert presents us with such an experiment that is occurring naturally.

That brings us back to our microflora. The action of water in previous geological periods, the erosion caused by the desert wind, and the nonbiological and biological dissolving that leads to the formation of metallic hydroxide crusts together constitute the chief weathering factors in the desert. These combined forces are capable of breaking down, exposing, and carrying away deeper and deeper slices of the earth's crust. That has decisive importance for historical geology. The inanimate forces of weathering and the activity of the microflora not only dissolve minerals that have already been oxidized—trivalent iron and manganese, for instance. They also bring to the surface oxidizable minerals, as well as some of the carbon that over the eons has been bound up in the rock of the crust. The minerals and carbon come into contact with atmospheric oxygen and are oxidized. Oxidation is constantly taking place in all the desert regions of the continents.

It is significant that in vast areas of the continents living conditions are already so poor that only microflora can continue to exist. Microflora, presumably much like those we know today, especially the lichens, began the settlement of the protocontinents some 450 million years ago, marking the end of the earth's primordial desert. The microflora ushered in the land phase of geological history. Today those same organisms represent what will surely be the last vegetative growth on continents that are steadily drying out. The symbiotic activity of the desert lichens is hardly effective for the production of free oxygen, since the oxygen that the algae produce by photosynthesis is consumed by the fungi. In the interior heartlands of the deserts, where there is not even enough moisture for the undemanding microflora, the inanimate forces of weathering continue to bare new and deeper strata of the earth's crust, exposing more oxidizable minerals and carbon to atmospheric oxygen. No layer of rock, no matter how deeply hidden, is immune from the mighty process of mountain formation—followed by elevation and weathering—at least once in the course of geological time. So great is the effectiveness of weathering that the earth's crust has a virtually infinite surface. We are made forcibly aware of this when we look at a weathering mountain in the desert. The pieces of rock that have broken off cliff walls form enormous heaps of scree extending all the way into the low-lying desert basins. The farther into the valley the talus is carried by gravity, the smaller the pieces of rock become, until at last it disintegrates into sand and dust. At the same time the rock's surface is constantly enlarged. A rock that has split in two has nearly doubled the surface it exposes to the air.

The infinite enlargement of the earth's surface by weathering can be seen very plainly from the example of a zeugen. Imagine a cube-shaped zeugen, each of whose sides measures 250 meters. Allowing for all the irregularities in its vertical walls and its tablelike top, it has a surface of about a square kilometer. In 100,000 years, or perhaps 300,000, depending on the hardness of the stone, the zeugen will have been completely worn away, disintegrated into sand. If the grains of sand have a diameter of 0.2 millimeter, the butte's area of a square kilometer will have become about equal to the land area of France. Such an example makes it clear how, in the course of the earth's future, all the remaining oxidizable minerals in the rocks of the desert will sooner or later come to the surface.

The grains of sand in the golden dunes of the deserts are covered with a reddish encrustation of metallic oxides just like that of the rock of the hammadas. To the extent that the sand grains are made up of minerals not yet saturated with oxygen, oxidation of the minerals is taking place on the surface of the grains. It is unlikely that the thin oxide coatings arise from the action of microflora; the friction is too great for them to take hold on windblown grains of sand.

The reduction in the vegetative production of oxygen on the continents because of the spread of deserts, conjoined with the persistent and increasing effects of oxidation, may one day result in a considerable diminution in the supply of atmospheric oxygen. An additional factor is energy-hungry man's constant consumption of oxygen by the burning of concentrated accumulations of fossil organic carboniferous materials in the form of coal, petroleum, and natural gas.

Photodissociation, the release of free atmospheric oxygen through the action of ultraviolet light in breaking down molecules of atmospheric water vapor, is not very productive, as we noted in an earlier chapter. The process would have produced no more than a fraction of the present oxygen content of the air. Also, without the contribution of lush plant growth on the continents, the oxygen released by the plant organisms of the oceans would almost certainly not be enough to maintain the present atmospheric level—the level that permits the respiration of highly evolved life forms, including man.

Some 450 million years after the first plants left the ocean and initiated the continental biophase, many regions have again become so arid and inhospitable that living organisms cannot survive in them. Such regions are again without plants to release oxygen into the atmosphere. From their readings of the evidence, many desertologists postulate three basic geological stages for the continents of our planet: primal desert, biophase, final desert. What we are observing today in the oxygen-consuming arid regions of the continents could well be the beginning of the final desert phase. It is possible that the final desert will not achieve total dominion over the continents until several billion years have passed—by which time humans would long have ceased to populate the planet. For the moment, we can only be struck by the tenacity of life against terrible odds.

Perhaps, 5 billion years from now, when the history of our earth has been completed, it will be apparent that the movement of living organisms out of the shelter of the oceans onto the continents was only a brief excursion.

# 6 / A Brief Cooling Off

## The Greatest Picture Gallery on Earth

If a line is drawn on a map of Africa from Alexandria to Dakar, and another from Addis Ababa to Marrakesh, the two lines will intersect roughly in the center of the Sahara. Near this point, in the heart of the sun-parched, waterless sandy waste, rise several narrow, tall rock pillars of reddish sandstone. They suggest the outstretched fingers of a gigantic stone hand buried in the sand.

One of those rocky fingers differs from the thousands of similar structures in the Sahara by the fact that about 8,000 years ago, or possibly as much as 12,000—there is a wide margin of error in the dating—a human being passed by. He or she picked up a stone that was harder than the reddish sandstone and began to draw on the surface of the rock, which had until then been marked only by the inanimate forces of weathering. And so a work of art came into being on that rock (whose age can be estimated at 400 million years)—a work of art that is perhaps the most beautiful and impressive rock carving on earth.

The unusually large composition shows three life-size cattle and a calf, their heads bowed over a water hole. The longer we look at the rock carving—reproduced in Plate 38—the more details the picture reveals. The two animals on the left are depicted so that their heads appear one above the other; the animal on the right is shown in profile. The lyre-shaped horns, the eyes, the ears, and the nostrils are carefully carved out and polished. In the animal on the right, even the spotting of the hide is represented. The same animal, which is

38. Rock carving, about 8,000 years old, showing a herd of cattle drinking. In the central Sahara.

242

drinking, has water running out of its mouth. In the lower left corner the head of the calf is visible; it has not yet grown horns.

Similar though more simply executed carvings, representing such wild animals as gazelles, rhinos, giraffes, and elephants, can be found throughout the Sahara (see Plate 39). While these art works are distributed over a vast area, approximately 20 miles north of the pillar of rock with the carving of cattle there is a kind of open-air museum consisting of thousands of colored rock paintings. The formations of the wildly fissured Tassili range, in the central Sahara, are in places bedecked with rock paintings. Here, among "cliff gardens,"

"avenues," narrow "streets," and broad "plazas," peoples of many different cultures, over thousands of years, covered the walls of cliffs and the ceilings of caves with paintings.

As one tramps through these mountains, it is hard to avoid thinking that painting must have been far and away the preferred occupation of the peoples who settled the Sahara thousands of years ago. In the number and variety of their paintings, they have left us an illustrated lexicon of their culture, their living conditions, and their environment. The paintings, like the rock carvings and drawings scratched in stone, repeatedly show animals that are still seen in the savannas of Africa. Even fish are depicted. So are the people and their lives. Some of the paintings are believed to be self-portraits of the artists; there are also graceful figures of women, mothers playing with their children, youths adorned with wreaths of flowers, depictions of religious ceremonies. I came upon one of the most beautiful of the paintings, a running archer, in a remote cave in the eastern part of the mountains (see Plate 40).

All these graceful figures have been painted on the roughest kind

of rock. Again and again cattle have been perfectly represented by the use of dots or triangles. There are life-size cows drawn with a single line, and large and densely packed herds whose motion is so well rendered that one can almost hear the trample of hoofs and the animals' low bellowing.

Plate 42 conveys an idea of the beauty of these rock paintings, some of them in polychrome, executed by Stone Age artists.

From the physique of the people depicted, as well as their tattooings and masks, it can be deduced that Negroid peoples with their herds of cattle probably migrated into the Sahara along with the animals of the savanna. The evidence of the paintings, when combined with Théodore Monod's discoveries of the bones of hippopotamuses and traces of fishing villages in dried lake beds in the heart of the desert, is conclusive proof that a much wetter and more fruitful climate prevailed in the Sahara between 8,000 and 10,000 years ago. The broad plains of the Sahara were peopled by gazelles, giraffes, zebras, and elephants, as the savannas of East Africa are today; hippos and crocodiles wallowed in countless lakes; and the high mountain ranges, from which many rivers sprang, were covered with forests.

It is fascinating to find that a few living relics have survived from this damp climatic period in the history of the Sahara. Where the streetlike canyons of the Tassili mountains open out into plazas filled with bizarre rock pillars, some gnarled trees still stand—no more than fifty or sixty of them—dating back to that remote epoch (see Plate 45). A few specimens of these mighty cypresses, whose seeds will not grow in the present climate of the desert, may be as much as 3,000 or 4,000 years old. Along with the sequoias of California, they are the oldest living organisms on earth. Thousands of years ago, as the climate of North Africa gradually turned more arid and rains fell more rarely and more sparsely, the trees slowly adjusted to the heat and aridity. They slowed their metabolism. But only the few specimens growing in the depths of the canyons have survived the climatic change; in those locations, a little water can collect during the occasional rains.

Some desert explorers claim to have found earthworms—those very symbols of a rainy climate—in isolated spots that have remained wet since that distant past. But the most interesting animal relics include

39. Rock carving of a rhinoceros in the northwestern Sahara

creatures hardly to be expected in the heart of an enormous desert: fish.

I had heard about the existence of these desert fish and have been able to verify the stories. Very small numbers of them do, indeed, live in tiny pools of water in the high mountains of the central Sahara (see Plate 48). These water-filled hollows in the granite are found in a region that was once the source of a river flowing south, probably into the Niger. As the desert spread into the lowlands, the river dried out, trapping the fish in its upper course. The fish retreated toward the river's source, a region that still receives some water because the sparse rains of the central Sahara tend to fall in the higher altitudes of the mountains. In a similar way, crocodiles were forced back on a mountain in the southeastern Sahara. The fish, isolated for thousands of years in the specific living conditions of a great desert, have evolved into a separate species—desert fish. There are comparable species of fish in the North American deserts.

The ancient cattle breeders probably suffered somewhat the same fate as the fish. When the desert began to spread over the plains, the people could only move with their herds to the cooler and moister higher elevations of the mountains. But the desert crept inexorably up the mountain slopes. The pastures for the cattle became parched, and their water holes dried out. That process undoubtedly went on for centuries, and it has been dramatically recorded in the rock paintings of the Tassili mountains. The more recent paintings often show scenes of battles between groups of warriors armed with spears or bows and arrows. Since herds of cattle can be seen in the background, the battles may have been part of a bloody struggle over grazing lands and water holes. In the end the desert was the victor. The cattle breeders were driven from the Sahara and retreated to the south. The remaining tribes became nomads. That fashion of life—a few people with small herds of livestock following the rare rains in search of grazing grounds—represents man's adjustment to the extreme living conditions of the desert.

The paintings of latest date in the Tassili mountains—probably done by tribes that were already nomads—show oases of date palms and camel caravans. Those pictures mark the end of a painting tradition thousands of years old.

40. Rock painting in the Tassili mountains in the central Sahara

## The Role of the Ice Ages

The fact that the Sahara region of North Africa was covered by savannas and forests as recently as 8,000 to 10,000 years ago has repeatedly led specialists, as well as laymen, to draw the false conclusion that the Sahara itself is only a few thousand years old—in other words, that it is a very young desert. That is no more true than the widespread fallacy that the vast desert was created by the activities of man, through deforestation and the destruction of the vegetable cover by overgrazing. Man is only contributing to its constant expansion—chiefly because of his herds of livestock.

The typical desert terrain seen by every traveler in the Sahara—the areas of monuments, the wind-worn Tassili mountains, the sandy oceans of dunes, the endless wastes of gravel in which everything has been worn down, leveled—could not have been formed in a few millennia. Millions of years were needed to produce such a landscape. As has already been shown, the deserts are an ancient phenomenon on this planet. We can tell from the stratifications of ancient rocks that the Sahara originated as far back as the Mesozoic era, about 150 million years ago.

At intervals of approximately 300 million years ice ages have occurred. Even at the end of the Carboniferous period, when the vegetative biomass was probably at its maximum, great areas of what are now the southern continents, and were then one coherent giant land mass, were covered by a tremendous armor of ice, as Antarctica is today. And in still earlier geological epochs, long before plants and animals left the seas to populate the land, the arid continents were subjected to extensive glaciation due to the cooling of the climate.

Every ice age is subdivided into periods of cold and interglacial periods of warmth, so that within a given glaciation large temperature fluctuations occurred at relatively short intervals. Although each of the ice ages (their causes are still largely unknown, but possible causes will be discussed in chapter 8) lasted several million years, together they constituted relatively minor interludes in the generally warm climate of our planet. The last great period of ice ages, about 2 million or 2.5 million years ago, was only a brief interruption in the drying out of the continent of Africa, which had been going on for more than 100 million years.

41. Looking for rock paintings among the bizarre forms of the Tassili mountains in the central Sahara

42. Rock painting, about 8,000 years old, showing a herd of cattle. In the Tassili mountains.

43. A nomad woman pounds hard desert grasses with a mallet to make them flexible. The tents in the background are woven out of these grasses.

44. Nomads on the move in the northwestern Sahara

45. A survivor of the last ice age: a cypress tree several thousand years old in the Tassili mountains of the central Sahara

During the cold phases of that ice age—the last one reached its maximum between 20,000 and 30,000 years ago—the polar icecap spread far to the south. Climatic zones likewise shifted south. With great parts of Europe, Asia, and North America coated in ice, the rain-bringing low pressure systems probably arose in the middle Atlantic, rather than in the northern Atlantic as they do today. And they crossed the Sahara region, instead of northern Europe. A green North Africa was the consequence. It was at the end of that last cold phase that many of the rock carvings and drawings were created, and the plant and animal relics of the Sahara also date back to that time. Immediately afterward, the desert began slowly but inexorably to spread again, in North Africa and in the other arid regions of the Northern Hemisphere. Vegetation gradually withered; the animals of the steppes withdrew or died out; and men were for the first time confronted with desert. The parching of North Africa took place slowly. As the present extension of the Sahara's southern boundary shows, the process is by no means complete.

The Sahara, then, did not arise in historical times. Up to the present, it has only reconquered the territory it lost temporarily during the last cold phase of the ice age, territory it had possessed for geological ages.

## The Illusion of a Great Vegetable Garden

Many technologists of the twentieth century seem to think it is only a question of time before man converts the deserts—chiefly the Sahara—into a vast vegetable garden. Most such dreamers, as I have often noted, have never seen the Sahara with their own eyes. To them, in an age of technology a desert is only a temporary problem. What is their optimism based on? As desertologists have increasingly turned their attention to the geological substructure of North Africa —explorations prompted largely by the quest for oil—they have discovered many sizable underground bodies of water. For many a nonspecialist, it has been only a small step from such a discovery to

46. Monuments in the Colorado Plateau in Arizona. Geologists call such formations zeugen. This picture is a clear rendering of the characteristic red coloration of the desert.

47. Nomad girls in the northwestern Sahara

the conviction that inexhaustible supplies of water lie beneath the Sahara. The Arabs call the Sahara *bahr belà mà,* "sea without water." Now the notion has arisen that it is a "sea with water."

That notion is fallacious. With some few exceptions, ground water occurs only where porous strata of rock are filled with water. But if the structure of the rock strata does not permit the ground water to flow away, the water may remain locked in the innumerable tiny spaces in the rock for several hundred million years.

The Sahara consists of nine vast geological basins in which thousands of feet of the most varied sediments have been deposited one on top of another over geological ages. These water-storing sedimentary basins are separated from one another by largely impervious thresholds of bedrock. The center of one of these enormous, bowllike sedimentary basins frequently is recognizable by its great fields of sand dunes, which represent the most recent filling the basin has received.

The basin sediments are mainly of two types: marine sediments, which were deposited during the long ages when the sea flooded large parts of North Africa, and continental deposits, from mass-wasting when the land was above water. It was chiefly salt water that was locked into the marine deposits; the continental rocks that were laid down during a wet climatic phase were largely filled with fresh water. In other words, geological phases are recorded in the rocks not only by fossilized plants and animals but also by fossil water. Porous sandstone of Mesozoic origin contains rain that fell from the sky 150 million years ago. Even in sedimentary rock laid down more than 400 million years ago, before the continents were occupied by plants and animals, primary water has been found. The ground water of the Sahara is largely fossil. It has been stored in the rock for inconceivably long ages.

In order for such ancient ground water to be preserved, the storage rocks must be buried deep beneath other, younger deposits, so that the water cannot flow away or evaporate. That applies to the storage of ground water in other desert regions besides the Sahara. But the statement is an enormous simplification. The geological structures and the presence or absence of fossil water are exceedingly complex matters, hard to analyze and in many areas of the world still awaiting

48. Fishing in the central Sahara. A few desert fish, isolated for thousands of years, still live in water holes such as these, fed by springs near the source of a largely dried-up river.

examination. Over geological ages, there can frequently be exchanges between widely different water horizons. If, for example, a continental sediment was deposited on top of a marine sediment, or vice versa, a mixing of fresh and salt water may take place; or there can be secondary fillings of storage systems that have been emptied for one reason or another. In the Sahara basins, the topmost strata of rock are in many places filled with the more recent waters of the last ice age. What is more, the character of the Sahara ground water is dependent not only on its origin but also on chemical processes determined by the widely varying nature of the storage rocks. If, for instance, secondary fresh water penetrates into rocks that were previously filled with salt water, the fresh water may be converted to salt by the deposits left behind in the pores of the rocks.

Although the store of water underneath the Sahara is enormous, only a small part of it is really suitable for irrigation projects. In addition, the water often lies hundreds or even thousands of feet below the surface, and unless it is under artesian pressure it cannot be brought to the surface. To apply an economic standard, water can profitably be pumped up from a depth of no more than 50 feet.

For irrigation projects, then, only storage rocks close to the surface and filled with ice-age water would lend themselves to exploitation. But wells and pumps installed by man encounter a tremendous rival for such water: the sun. The sun operates like a gigantic pump that literally sucks water out of the underground reservoirs. The average annual evaporation in the Sahara is a water column measuring about 20 feet. In some cases a world record for annual evaporation has been recorded: 25.6 feet. In some areas the ground water lost by evaporation apparently is replaced from torrential downpours; but in other areas the ground water level is rapidly sinking, for it has not been replenished since the last wet period in North Africa.

As anyone can testify who has ever spent any length of time in an oasis, the principal enemy of irrigation farming is salinity. Evaporation is so great—it continues even at night—that vast quantities of water must constantly be pumped into the fields and date palm plantations. But only a very small amount of this water actually benefits the plants. Most of it evaporates, and even water with few minerals dissolved in it leaves behind a thick encrustation of salt. Also, the feverish search for petroleum, which has turned up so many new sources of water in the Sahara as a more or less accidental by-product, has prompted young workers to leave the oases for the high wages to be earned in the oil fields. The springs and irrigation systems of the oases, fashioned over the years on the basis of an intimate knowledge of ground water conditions, have been drying up and becoming choked with sand. Over large areas of the Sahara, the domain of irrigated cultivation is shrinking, rather than increasing.

The application of technology by inexpert hands can have devastating effects. The development projects on the southern margin of the Sahara are a prime example. When extensive supplies of ground water were discovered in the region, hundreds of new wells were drilled. As a result, the herds of cattle, the pride of the local nomad tribes, multiplied enormously. Overgrazing of the sparse vegetation by the bigger herds led to the spread of the desert. Within a short time the meager pastures were totally destroyed and three-quarters of the cattle herds were dead of starvation. For many years previously, the equilibrium between the supply of food and the number of animals had been relatively stable. The intervention of technology had done irreparable harm by shifting the balance in favor of the desert.

Even given optimum utilization of the Saharan ground water, with careful ecological controls, only short-term successes seem to me to be possible. For the supplies of largely fossil ground water can be exhausted. In addition, the irrigation technologists must reckon with another set of difficulties and dangers. As alarming reports from the American deserts indicate, any long-term withdrawal of ground water from coastal desert regions—which are the best for irrigation projects—leads to such a lowering of the ground water level that sea water from the edge of the continent penetrates into the storage rocks. I myself saw sea water, rather than fresh water, come pouring out of the pumps one day, transforming what had been fruitful fields and orchards into a white, sterile salt desert.

Another option that has been considered for "making the desert bloom" is the desalinization of sea water—which certainly is available in unlimited quantities. Some experimental plants of this kind are presently operating in several sheikdoms on the Persian Gulf. Unfortunately, both their initial cost and their consumption of energy are so high that a pound of tomatoes produced by irrigation with desalinated water would have to be priced at $9. Yet it is strange that the oil-rich desert countries do not use their precious raw material primarily to develop the technology of irrigation. Instead, they are shortsightedly selling most of their oil to industrial nations, so that the latter can maintain their excessively high standard of living—much of which is sheer waste. In any case, that living standard is bound to collapse in the foreseeable future, when the supplies of oil are finally exhausted.

In this connection, the hope of someday being able to utilize atomic energy to desalinate sea water seems to me exceedingly naïve. We should have learned from history that human beings are more likely to use this source of energy to further the goals of power politics, not for irrigation.

But even if there were meaningful collaboration between countries wealthy in raw materials and countries with highly developed technologies, it would still be sheer fantasy to try to turn the entire Sahara into a vast market garden. A close knowledge of local conditions tells us that, at most, perhaps 1 per cent of it can be irrigated on any lasting basis. The successes of the Israelis with irrigation are repeatedly cited, and cannot be ignored; but, as a glance at the map will show, they amount to no more than a drop of water in the hot sand of the desert.

Recent developments in the Sahara have unleashed successive waves of irrigation euphoria. Yet the desert is spreading by some 40 square miles every day. That is as much in a day as is likely to be reconquered by irrigation in a whole year. In the future, it will undoubtedly prove more "fruitful" to limit the number of human beings by population planning than to irrigate the deserts.

# 7 / Cosmic Models

## Neighbors in Space

Human beings probably observed the heavens and noted events in them hundreds of thousands of years ago. Of course, prehistoric man did not recognize a solar system or know anything about celestial mechanics. He could grasp only what immediately affected his life. The daily recurrence of the sunrise poured out warmth upon nomadic tribes that had shivered in caves through the night, and the light of the sun enabled the hunter with his stone ax to track his quarry.

As the human capacity for thought developed, some searching minds began to reflect on the meaning of the events they observed in the sky. Why did the sun rise in different places? Why was it low over the horizon during the cold winter but straight above their heads in the summer? Why did the moon constantly change its shape? Did such observations have any meaning for men?

Somewhat later in mankind's cultural evolution, observation of the heavens became one of the primary fields of knowledge among many peoples. And the study of the sky gradually altered human consciousness. The observations and calculations of astronomers explained the causes and effects of natural phenomena and laid the cornerstone for objective science. It turned out that the sun did not stand low over the horizon and radiate a weaker light because winter was cold, but that it was winter because the sun's rays struck the earth's surface at an oblique angle. After millennia of observations, astronomers developed a calendar and bestowed a sense of space and time on humanity. For the first time in the long history of the earth, it became possible for a living organism to experience its existence temporally as well as spatially within the cosmos. That expansion of consciousness was later to form the basis for one of the most important of scientific discoveries.

Today, microbiologists, with the help of the electron microscope, try to penetrate the secrets of the smallest building blocks of life, and astrophysicists, sitting at the eyepieces of giant telescopes or the switchboards of enormous radio telescopes, are trying to fathom the evolution of life and the past and future of the universe. Man, driven by his passion for knowledge, has employed his technology to leave the earth itself and set foot on another heavenly body. For the first time, men have stood on the moon and looked at their own planet moving through the cosmos, as in the past they looked up from earth at the other planets of our solar system and at the stars. That objective view has produced a new appreciation of the situation of our home planet; it has forced upon us a comparison between the earth and neighboring planets. Astronomers are insistent about it: we can understand the earth only if we know her sister planets. The statement is particularly true for desertology. Desert scientists start from a simple, practical consideration. If they could investigate another planet that was in either a very early or a very advanced state of its evolution, useful conclusions might be drawn about the distant past or the distant future of the earth. Above all, observation of other planets might tell us whether we can expect the earth's course to run from primal desert through biophase to a final desert.

But is it possible to compare other planets with our own? When we consider the number of unsolved problems presented by the physical nature of our own planet, and how difficult it would be to obtain from another planet even a fraction of the scientific data we have about earth, such an undertaking seems hopeless. Besides, we have no reason to believe that any other planet has gone through a shorter or longer evolution than earth's. All the planets in our solar system consist of the same elements, and all came into being at approximately the same time, about 5 billion years ago. No planet in our system is significantly younger or older than its neighboring planets. Nevertheless, the evolution of each has taken a different course, and that very difference can provide information about the past and future of the earth.

The physical evolution of a planet, and especially the appearance of life on it, is not just a matter of time. It is affected by two other primary factors: the planet's distance from the energy-giving central star (which determines its surface temperature) and the size, or, more precisely, the mass, of the planet (which determines its gravity).

Remember that a complex set of physical conditions is necessary for life to arise on a planet. One absolute prerequisite for the origin of life is, as we have seen, the presence of water, primarily in liquid form. Billions of years ago, the first building blocks of life and the first organisms developed within the shelter of water. Water is the universal solvent. Almost half the known elements are soluble in the earth's waters. Without the oxygen dissolved in water, fish could not breathe in the oceans, and without water to serve as a solvent nothing could be fed. The roots of plants can take nutrients from the soil only in soluble form, and man's food must also be dissolved before it enters the blood circulation. The most important processes of life on our planet, such as plant photosynthesis, require water. In short, no living process would be possible without water in its fluid state. In addition, water must be present in the gaseous state. Water on the surface of the oceans is heated by the sun, rises as a gas into the atmosphere, and condenses again in the clouds. Only in the form of rain-giving clouds can the vital liquid be transported from the basins of the seas to the higher altitudes of the continents.

In order to guarantee the presence of water in solid, liquid, and gaseous forms, our planet must have a specific size and must orbit the sun at a specific distance. If, for example, the earth's orbit lay closer to the sun, the surface temperature would be so high that water would be present only in gaseous form. Moreover, the high ultraviolet radiation would probably have split all the molecules of water into hydrogen and oxygen, and no water emerging from the planet's interior could have lingered on the surface. While the oxygen would be consumed by the rock, the hydrogen, because of the high temperature, might well leave the gravity of the earth altogether and escape into space. On the other hand, if our earth orbited the sun at too great a distance, the temperature of its surface would be so low that water would be present there only in solid, frozen form.

The size and mass of a planet are also crucial factors. On a small planet of low mass, the atmosphere would escape from the low gravity after a relatively brief initial phase and would dissipate into space. If such a planet had started on a biological evolution, that phase would end much sooner than on earth, before highly evolved organisms could develop. The earth is still in an animate phase, but a small planet would probably have reached its final phase long ago.

On such a planet we could observe and study a kind of earthly future.

On a planet of large mass, a biologically determined evolution and the transformation of the protoatmosphere into a life-supporting, oxygen-rich atmosphere would probably never begin. Because of the high gravity and the large amounts of gas vented by its huge body, the planet would have a protoatmosphere of enormously high density and pressure, compared with that of earth. The high pressure would make the protoatmosphere, consisting chiefly of reducing gases, very different chemically from the protoatmosphere of earth. In addition, it would probably be far too dense to admit enough sunlight for plants to embark on the oxygen-liberating process of photosynthesis and so create an atmosphere comparable to earth's.

Evolution on a planet with a greater mass than earth's would come to a halt early in its atmospheric development, and its surface would probably never develop beyond the stage of primal desert.

The gravity of earth is precisely enough to retain a life-supporting, sunlit atmosphere consisting of water vapor and other gases; and the planet's distance from the sun of 92.9 million miles keeps it in the narrow zone where water can exist as ice, as a liquid, and as a vapor.

The physical facts, rather than the duration of time, are what primarily determine the development of a planet. In drawing conclusions about our own planet from observations of other planets, we assume that the same physical and chemical laws apply everywhere in the cosmos, an assumption that indeed seems likely. Our galaxy contains about 100 billion suns, of which about 5 billion, according to recent estimates, may have planetary systems comparable to our solar system. But unmanned space probes—not to mention manned space flights—cannot overcome the enormous distances between stars. So our observations are limited to our immediate neighbors in space, Venus and Mars, which are among the inner planets of the solar system. In fact, these two are the sole planets whose history can be compared with that of earth, because they are the only ones comparable in size and in distance from the sun.

Mercury, which is also one of the inner planets, is a special case because of its low mass and its closeness to the sun. Conditions on that planet certainly cannot bear comparison with those on earth. The same is true of all the outer planets. Pluto cannot be compared because it orbits our central star at such a great distance that the sun's rays cast only a feeble light on its surface. The other outer planets are so massive, besides being so distant from the sun, that their evolution cannot even remotely be compared with that of earth (see the drawing on page 48).

In many respects, earth's nearest neighbor, Venus, offers the closest parallel. Venus is almost a twin of earth in size. Its equatorial diameter is nearly the same as earth's (7,830 miles, as against 7,922), and its mass amounts to 81.4 per cent of earth's. But any study of the surface of Venus by telescope is virtually ruled out by the density of the planet's atmosphere. As early as 1932, spectroscopic studies of the light reflected from Venus showed that the atmosphere contains large amounts of carbon dioxide. During the past two decades, studies by radio telescope and by unmanned space probes have given us a reasonably exact picture of the planet's physical and chemical make-up.

The picture that scientists have pieced together from countless bits of information shows a tremendously hot desert planet lashed by sandstorms. The picture has probably been much the same for the past 5 billion years. The surface of the planet turned toward the sun has temperatures that range between 400° and 700° C. An earthly observer would behold a fearfully apocalyptic desert landscape, compared to which the hottest and most inhospitable regions of the earthly deserts would look idyllic. A thick, gloomy layer of clouds casts an eternal twilight on the sand-lashed desert surface of Venus, and the rocks that wind and sand have been deforming for millions of years glow a dull red in the shadows.

The cause of these infernal conditions is the dense atmosphere surrounding the planet. It consists of about 95 per cent carbon dioxide, and it is so thick that the pressure on the barren surface of Venus amounts to seventy-five earth atmospheres. The "greenhouse effect" traps the sunlight that falls on the surface of Venus and results in an enormous heating of the planet.

Data from the Mariner 2 space probe have told us much about the planet's development. Venus early failed to form a hydrosphere because its closeness to the sun kept the surface temperature too high for water vapor to condense. Most of the water vapor molecules vented by the rocks were soon split by the strong ultraviolet radiation, and because of the high temperature the light hydrogen easily escaped the planet's gravity and was lost in space.

At the same early stage on earth, on the other hand, surface temperatures were low enough for the water vapor to condense and form a hydrosphere, which eventually was able to absorb almost all the carbon dioxide component of the atmosphere. Venus could not similarly dispose of its increasing carbon dioxide, and the rising temperature created by the greenhouse effect only made the situation worse. Consequently, Venus never developed beyond the stage of primal desert. To this day, it presents a picture roughly corresponding to that of the desert earth some 4.5 billion years ago.

## The Secret of the Red Planet

For thousands of years, the "red planet" of Mars has stirred men's imaginations more than any other in the solar system. Because of its reddish glow, it was considered to be a heavenly symbol of the blood. Its dim red gleam was also associated with death and misfortune. For that reason, the ancients assigned it the name of their war god and performed human sacrifices to appease its wrath. Later, especially after the invention of the telescope, Mars became a great favorite with astronomers. Its two moons were given the names of the war god's servants, Phobos and Deimos—Fear and Terror.

Mars resembles the earth less than Venus does with respect to size and mass. Its diameter of 4,200 miles is just a little more than half the earth's diameter. The planet's mass amounts to a tenth of the earth's, and its total surface is about equal to the area of our continents alone. The planet's mean distance from the sun, which it orbits in 687 days, is approximately 142 million miles, more than half again the distance of the earth from the sun. On the average, the sunlight on Mars is about half as intense as it is on earth.

The planet has attracted great attention from scientists mainly because its surface is not, like the surface of Venus, hidden beneath an extremely dense atmosphere. It can fairly be called the most thoroughly studied planet in the solar system—except for our own planet, of course. Its clear, almost cloudless atmosphere and its prevailing red color caused it to be quickly recognized as a desert planet. Mars, then, provides a model for the evolution of deserts, a model that may permit us to read the future of our earth.

The atmosphere of Mars appears to be extensive enough and thick enough for the planet to have something like climate and weather in our sense. The inclination of the axis of rotation is 25 degrees, almost exactly that of earth, so that the northern and southern hemispheres of Mars are alternately turned toward and away from the sun during the planet's annual revolution. But that revolution takes 687 days, making a Martian year almost as long as two earth years. Correspondingly, the seasons are twice as long as on earth. The length of the Martian day, however, is very close to ours: 24 hours and 37 minutes.

Mars has white polar caps, as earth does, and in many places the coloration of the surface changes with the change of seasons. It can be compared with the earth in many respects, and as long as human beings have been observing this neighboring planet they have been tempted to speculate that life comparable to earthly life might exist there.

Searching for signs of life on Mars by purely visual means is extremely difficult. Although its atmosphere is usually cloudless, observation from earth is attended by great problems. Because of the turbulence of the earth's own atmosphere, even the strongest of telescopes show Mars as merely a small, blurred red sphere. An unpracticed observer may make out nothing but the gleaming white polar caps and a few dark and bright spots and lines. With practice and patience, as well as good viewing conditions, the blurred reddish sphere begins to reveal remarkable complexities. Finer features and colored areas become visible. The observer thinks he can make out the dim outlines of continents, seas, and cloud cover. Gradually the planet seems to fall into geographic patterns familiar to him from those of earth. If the observer is lucky and happens on one of those rare moments when the air above the telescope is completely still, the blurred sphere suddenly appears sharp and clear, presenting hundreds of fascinating details that cannot be set down fast enough in a hastily drawn sketch. For suddenly the vision vanishes again, like a wraith.

Until 1965, when the Mariner probes of Mars began to report their sensational findings, descriptions of the desert planet varied from observer to observer. The data, the sketches, and even the photographs astronomers took of the Martian surface frequently disagreed. From these contradictions arose the canal controversy, one of the most curious incidents in the history of modern astronomy.

In 1877, the Italian astronomer Giovanni Schiaparelli announced that he had discovered thin black lines on Mars that covered the planet like a spider's web. He called the lines *canali*, "channels" or "trenches." It should be noted that Schiaparelli never claimed that his *canali* were artificial canals built to carry water. In fact, he assumed that they were geological trenches or faults.

Map of Mars by Schiaparelli

The American astronomer Percival Lowell, one of the leading specialists on Mars of his day, was excited by his Italian colleague's observation. The longer Lowell studied Mars, the more "canals" he discovered. Soon he came forth with the fantastic hypothesis that the network of lines represented canals dug by the highly civilized inhabitants of Mars in order to irrigate the parching surface of their planet with melting water from the polar icecaps. The trouble with this hypothesis was that, even if the irrigation canals were several miles wide, they could not have been seen from as far away as earth. Lowell therefore concluded that wide strips of irrigated land lined both sides of the canals, forming the dark lines he saw. Where the canals intersected, Lowell speculated, there must be great cities.

As might have been expected, Lowell's hypothesis quickly found a multitude of enthusiastic adherents, whose imaginations eventually provided Mars with everything a planet populated by civilized beings might need, including cities built beneath glass domes for protection against sandstorms and insufficient oxygen. The cities, located, as Lowell had suggested, at the intersections of the irrigation canals, had mighty power plants to keep the water moving and to pump it out to the fields. Ships busily plied the canals connecting the Martian cities. From there, it was only a small step for Lowell's contemporaries to equip the newly discovered men on Mars with spaceships, to project the character traits of earthly humans onto the

Martians' oversized brains, and to send them on a voyage of conquest to earth. The panic created in the United States by Orson Welles's famous "War of the Worlds" broadcast—a radio play that began with a supposed news broadcast about Martian landings on earth—arose out of a combination of such imaginings and the war jitters of 1938.

Even the two moons that circle the red planet, Phobos and Deimos, were fitted into the speculations about possible intelligent life on Mars. Their small size and their closeness to the planet make the two moons of Mars unique among the known moons in our solar system. Deimos is the smallest of the known moons; its diameter is only about 3 miles, its distance from the surface of Mars about 15,000 miles. Phobos, with a diameter of 5 miles, is also tiny, and is only 5,600 miles from the planet. By way of comparison, our moon orbits the earth at a distance varying from 220,000 to 250,000 miles. Careful observations and calculations have shown that Phobos's period of rotation around Mars is slowly growing shorter. The only plausible explanation for such a phenomenon is that the outer reaches of the Martian atmosphere are gradually acting as a brake on the moon—which has the paradoxical effect of causing it to drop into a lower, shorter, and therefore faster orbit. But the planet's atmosphere is so thin that it could not influence a satellite in this way unless the satellite had an exceedingly low density. If the calculations are correct, Phobos must be lighter than any known solid, even lighter than naturally occurring porous materials.

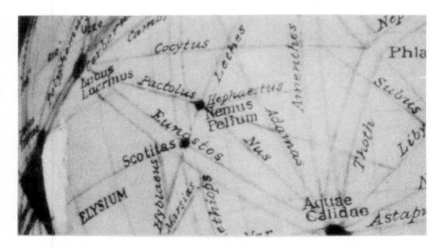

Section of Lowell's map of Mars

The Russian astrophysicist Joseph S. Shkolovsky drew some interesting conclusions from these perplexing facts. He raised the question whether Phobos might not be a hollow body with a hard shell. Perhaps, he suggested, Phobos and the smaller moon, Deimos, were huge artificial satellites placed in orbit by the highly civilized inhabitants of Mars several hundred million years ago. As the surface of Mars increasingly turned to desert, as the oxygen in the atmosphere was slowly consumed and the planet became uninhabitable, the Martians built the two gigantic satellites as libraries and museums to preserve the achievements of their advanced, but doomed, civilization. Maybe, Shkolovsky speculated, they hoped that someday these repositories would be visited by intelligent space travelers from other worlds. That intriguing idea had to be abandoned when the Mariner probes began their direct study of Mars. In the photographs they radioed back to earth, the moons of Mars revealed their true nature at last. They are shapeless rocks whose surface is scarred by countless meteorite impacts.

The mysterious irrigation canals underwent a similar deromanticization some years ago. Many of Lowell's colleagues had rejected the canal hypothesis as the product of an overexcited imagination. However hard and long they looked through their telescopes, they could make out nothing more than vague light and dark patches on the surface of Mars, although under especially good viewing conditions many dotlike details could also be discerned. Then the Greek astronomer Eugenios M. Antoniadi, working in France with an excellent telescope, had the good luck to observe, during a brief period of excellent viewing conditions, the same area in which Schiaparelli had discovered the mysterious lines. Antoniadi perceived that what appeared to be the thin lines of the *canali* were in reality a discontinuous series of irregular marks and dots that had probably looked like connected lines because of a common optical illusion. Like his colleagues, Antoniadi also noted large and rather oblong dark zones, but these were much too wide to be interpreted as canals.

In more recent decades, as the techniques of observation, measurement, and evaluation improved, the fantasies about canals and advanced civilizations gradually subsided. Where the potentialities of optical telescopes ended, work with radio telescopes began. Distant planets were mapped with the aid of radar beams. And in 1965, when the Mariner 4 space probe passed Mars at a distance of only 6,000 miles, the close-up study of Mars began. The probe sent back data on the composition and density of the Martian atmosphere, on sur-

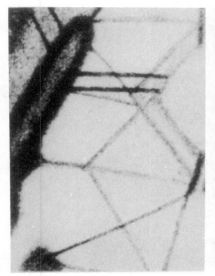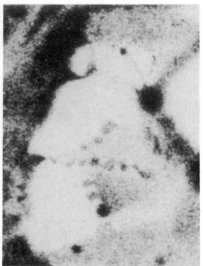

*(Left)* Cross-section of a map of Mars by Schiaparelli. *(Right)* Antoniadi drew an entirely different picture of the same region.

face temperatures and magnetic field intensities. It radioed back the first photographs of the surface of Mars, each photograph composed of 22,000 individual dots. On May 30, 1971, the Mariner 9 space probe was rocketed toward Mars. After a space flight of 167 days, over a distance of 250 million miles, the unmanned probe was placed in orbit around the red planet. In a year and a half of unique scientific accomplishment, the probe radioed more than 50 billion measurement impulses back to earth. Some of the signals, when put together, made up 7,300 spectacular pictures showing the entire surface of the planet. The photographs are so clear and sharp that exceedingly small details on the Martian surface can be seen. From them, and from the physical and chemical analyses that Mariner 9 sent back to earth, proof was finally obtained that Mars is not a completely dead planet, but is still being shaped by dynamic forces. Its surface is richer in detail than even Lowell had supposed, and its variegated topography can be compared to that of earth. Volcanoes up to 78,000 feet high loom above the plains—more than two and a half times higher than the highest mountains on earth. Vast red deserts are broken up by mountain ranges and upwarped cliffs. Canyons up to 2,000 miles long and 23,000 feet deep—almost four times as deep as the Grand Canyon of Arizona—cut through rocky wastelands on high plateaus as big as continents.

The new information has led to fruitful exchanges between terrestrial and Martian desertology. Mars has three principal surface forms. Basin-shaped lowlands cover some two-thirds of the planet; in size and shape they give the impression of dried-out ocean basins. As had long been surmised, they are filled with red desert sand and dust, and are what gives the planet its characteristic red color. The desert basins can be compared to the great lowlands of the Sahara, also filled with sand and dust. Interspersed among the basins are dark regions, the highlands. They occupy large areas, especially in the southern hemisphere, and give the impression of being coherent and vast continental masses. But the most striking feature of all is the planet's polar caps, whose glowing white contrasts so sharply with the color of the adjacent regions. The white polar caps shrink considerably during the summer and expand again in winter.

With each change of seasons, the dark highlands of Mars show changes of color that have long intrigued scientists. Similar seasonal changes are familiar on earth and can be observed from space, chiefly in the Northern Hemisphere. At the beginning of spring, the blanket of snow over Europe, Asia, and North America shrinks and retreats northward, until only the eternal ice of the polar region is left. The melting snow exposes the darker crustal rock of the earth. A few weeks later the land that has been liberated from snow again changes color as the trees put on new leaves, in a steady progression from south to north, and below the tree line the ground vegetation turns green. An intelligent living being on a distant planet, observing earth through a telescope during that period, would notice two successive waves of darkening running from south to north.

Something similar can be observed on Mars every year, but in the reverse direction. As one of the polar caps shrinks during the Martian spring, a wave of darkening proceeds from it and moves toward the equator at a speed of more than 25 miles a day. Curiously, the darkening occurs only on the already dark highlands, making them appear almost black. The low-lying red deserts, on the other hand, barely change color. Some scientists have propounded the interesting theory that the color change is caused by the sprouting of spring vegetation. They explain the "wrong direction" of the development of Martian vegetation by the argument that on an almost cloudless desert planet only the melting icecaps can supply the water indispensable for plant growth.

Tempting though it was to think that the earth might not be the only green planet in our solar system, the idea has been rejected by other scientists. They argue that the thin Martian atmosphere and the resultant low atmospheric pressure make it inconceivable for water to be present in liquid form.

The Mariner 4 space probe showed that the atmosphere of Mars is even thinner than had previously been calculated from measurements made on earth. As the probe flew past Mars and slowly disappeared behind the planet, its radio signals passed through more and more of the Martian atmosphere. From the weakening of the signals, astrophysicists were able to calculate that the planet's atmosphere is as thin as the atmosphere of earth is at an altitude of about 18 miles, and that the pressure it exerts on the surface of Mars is about 1 per cent of the earth's atmospheric pressure at sea level.

Unlike the earth's atmosphere, which is made up chiefly of nitrogen, oxygen, and about 0.03 per cent carbon dioxide, the Martian atmosphere consists almost entirely of carbon dioxide. Since it has only traces of oxygen and ozone, the ultraviolet rays of the sun, so deadly to earthly organisms, can reach the surface unhindered.

Certain types of clouds that occasionally appear in the Martian atmosphere evidently contain small amounts of water vapor, but the concentration is never as high as it is above the driest desert regions on earth. Because of its low density, the Martian atmosphere comes nowhere near being able to reproduce the greenhouse effect of the terrestrial atmosphere that allows it to store a great deal of the sun's heat. If one further considers the extreme dryness of the completely waterless Martian deserts, it becomes evident that the warmth that does reach the planet from the sun easily and promptly escapes into space. For that reason, the surface of Mars is primarily a cold desert. At the equator, the temperature by day rises at most to 20° or 30° C. (68° to 86° F.), but at night, because of the high radiation, it drops back to −85° C. (−123° F.). At the poles, winter temperatures actually drop to −130° C. (−202° F.).

Whereas the mean surface temperature on earth is 14.3° C., the mean temperature on Mars is probably −15° C. The thin layer of carbon dioxide that surrounds the planet just suffices to maintain a cool to very cold climate and to prevent nighttime temperatures from dropping down to the cold of space. The low temperatures, along with the extremely low pressure of the Martian atmosphere, make it unthinkable that water in liquid form could exist on the planet's surface.

The polar regions of Mars have such extremely low temperatures that they act as cold traps. They withdraw from the atmosphere its already sparse supply of water vapor. Most scientists agree that the polar caps consist partly of frozen water vapor that has passed directly from the gaseous to the solid state without an intervening condensation into liquid form, much like the frost flowers that form in winter on the inside of a cold windowpane. The reverse process also takes place: because of the low atmospheric pressure, ice sublimates directly into vapor, like dry ice, which passes directly into gaseous carbon dioxide when it melts.

If there is any sort of annual spring growth of vegetation in the Martian deserts—which is hardly believable because of the ultraviolet radiation—it could not satisfy its moisture requirements from water pouring out of the melting icecaps. Like the highly specialized hairnet plants of Death Valley, described earlier, the Martian plants would have to extract water vapor from the atmosphere. And even water vapor is probably not available in sufficient quantity during the Martian spring.

At the poles, not only sparse water vapor freezes into ice. The carbon dioxide that is the prevailing gas in the Martian atmosphere also freezes, becoming dry ice in the winter polar temperature of $-130°$ C. In all probability, the white polar regions consist not of frozen water but of frozen carbon dioxide. When the temperature rises in early spring, tremendous amounts of carbon dioxide evaporate, but most likely the temperature is not high enough for the frozen water to thaw or sublime.

During the summer, a mushroom-shaped cloud of bluish vapor can be observed over the poles, beneath which a white residue of ice remains. The ice is some hundreds of feet thick and probably consists of water vapor that has remained frozen.

Some of the most interesting photographs radioed back to earth from Mariner 9 were those taken during the Martian southern hemisphere's summer, which show the edge of the south polar cap after it had shrunk from 3,200,000 square miles to 32,000 square miles. The canyonlike slashes through the heart of the pole, along with the stratified deposits and dunelike shapes on the exposed ground around the edges, clearly reveal that the ice armor is built up of many different strata, like a layer cake. The coloration suggests alternating strata of water ice, volcanic ashes, dry ice, and desert dust.

As soon as I saw those photographs of the Martian south polar region, I immediately recalled having observed similar stratifications on earth. They were the alternations of snow and sand in southern

Colorado, described at the beginning of chapter 5. The deposition of various materials in the south polar area of Mars is excellent evidence of dynamic atmospheric conditions. In spite of the thinness of the atmosphere, strong winds are present, which transport water vapor, ash, and desert dust to the polar regions and deposit them in successive layers. Indeed, from visual observation, spectroscopic studies, and the polarization of sunlight reflected from Mars, it had long been known that violent seasonal sand and dust storms raged on Mars.

When Mariner 9 began its voyage to Mars in 1971, the atmosphere of the planet seemed fairly quiet. But when the braking rockets were ignited, after five and a half months of flight, and the space probe went into its predetermined orbit, it looked down upon a planet in turmoil. The first radiophotos that Mariner 9 sent back astonished the scientists. Mars was veiled by a sand and dust storm covering the entire surface of the planet. The mantle of dust was so thick that nothing could be seen but the white south polar cap and the peaks of four giant volcanoes, which thrust up through the masses of dust like islands surrounded by a raging red sea. It was the heaviest sandstorm that had ever been observed on Mars, and it did not subside until the beginning of 1972. When the storm ceased, the probe radioed back photographs that showed the planet with a clarity never experienced before.

The inhospitable conditions on Mars make it extremely unlikely that the deserts are covered by a blanket of vegetation every spring. A much more plausible explanation for the annual waves of darkness on the Martian surface is the storms. Reddish, powdery dust and very fine sand cover almost all the surface of Mars. This substance fills the deep desert basins and blankets the stony wastes of the black highlands with a thin film. That is the conclusion scientists have come to from their study of the polarization of sunlight reflected from the highlands. Changes in the light make it possible to estimate the size of the particles reflecting that light. It is now clear that the waves of darkening that proceed from the poles toward the equator every spring are accompanied by changes in the size of the dust and sand particles covering the dark highlands. When the strong polar winds prevail on Mars—they arise from the difference in temperature between the equatorial zone and the poles—the winds sweep the fine dust down from the highlands and deposit it in the huge desert basins.

Because of the thin atmosphere, the winds would have to reach velocities of about 95 miles an hour in order to lift and carry the dust, even though it is as fine as flour. The Martian winds would probably not be able to transport any of the coarser products of weathering, such as grains of sand, from the highlands to the lowlands.

In late summer, when the black highlands have warmed up, powerful updrafts form above them. Strong winds from the desert basins pour into the resultant low-pressure zones. These winds carry the fine dust in the reverse direction, from the basins back to the highlands, which then regain the lighter color that is characteristic of them in the autumn and winter.

A similar process can be observed in the deserts on earth. In the Sahara, the wind constantly transports the products of weathering—the sand and dust—from the dark, high rocky deserts into the lowlands, where it deposits the materials in the form of reddish sand dunes. Strong seasonal winds reverse the direction for a while, redepositing dust and sand from the lowlands onto the higher elevations, which temporarily show a lighter coloration.

One of the most interesting features of Mars that the Mariner photographs revealed is the many meteorite craters covering the entire surface of the planet. Most of the 12,000 or so sizable Martian craters are round, shallow depressions; unlike the meteorite craters on the moon, they do not have thick, bulging rims. Apparently they have been filled with red desert dust up to the level of the surface around them. The craters in the highlands look weathered; those in the lowlands are almost buried in desert dust.

At first glance, the surface of Mars, with its many meteorite scars, looks much like the surface of our moon. It is known that a great part of the present surface of the moon has remained virtually unchanged for 3.5 billion to 4.5 billion years. Probably because of its low mass, the moon never vented a sizable protoatmosphere—so that it made no difference that its gravity was too low to have kept the gases from escaping into space. The moon's surface has never been subject to erosion by wind or water, and so it is well preserved. That does not mean that the surface has undergone no changes at all. The proton radiation from the sun strikes the moon with full force, since there is no intervening atmosphere, and this "solar wind," as it is called, produces temperature differences that can crack rock. Moreover, meteorite impacts have shattered many of the moon's rocks. But such events have not brought about the total attrition of the moon's original surface.

Since the surface of Mars, like that of the moon, is pocked with meteorite craters, it is a good assumption that the planet is the same age as the moon. It can also be assumed that highly effective erosion such as occurs in a denser atmosphere or when surface water is present has never taken place on Mars. An animate phase in the planet's evolution would therefore seem unlikely.

Nevertheless, after evaluating all the measurements and photographs, Martian specialists have come to a different conclusion about the degree of Martian erosion, and about the possibility of life on the planet as well. Some special conditions arise out of the presence of the red planet's somewhat unpleasant neighbors.

Between the orbit of Mars and that of Jupiter there extends a broad ring of planetoids. This asteroid belt, as it is often called, contains between 40,000 and 60,000 small heavenly bodies—though they are big enough to be detected by telescope—which pursue their own orbit around the sun (see page 48). Many astronomers believe that the innumerable fragments of stone and metal in the asteroid belt might have formed another large planet if they had not been disturbed by the gravitational effect of the neighboring giant planet of Jupiter.

Among the visible planetoids are scattered several tens of millions of meteorites. They are so small that they cannot be made out as single objects, but they are big enough to gouge out huge craters when they wander from their orbit and strike a planet. Astronomers have calculated that Mars must have been struck by approximately thirty times as many meteorites as the traces on its surface indicate. In other words, if the surface of Mars were actually very old and without significant traces of erosion, it would be much more scarred than it is. In fact, it should show more meteorite craters than the surface of the moon. Since the transmitted photographs indicate that this is not the case, erosion must have operated very effectively on Mars, and the planet's present visible surface must be much younger than the surface of the moon. Astronomers have calculated that the surface forms on the planet as seen today are less than 600 million years old—compared to the planet's total age of about 5 billion years. Yet the changes and movements possible in the present thin atmosphere of Mars are not sufficient to have shaped the existing surface picture of the planet. Something else must have repeatedly reshaped the surface and worn away the craters.

Many indications in the data compiled from the photos and measurements radioed back by Mariner 9 do indeed suggest that Mars once had a much denser atmosphere and that water as well as strong winds shaped its surface, among other things eroding the old craters that date back to the period of the planet's formation.

Meteorite craters are not a phenomenon of the moon and Mars alone; several dozen of them have been found on earth. The best known and best preserved is in the Arizona desert and is more than three-fourths of a mile in diameter and 570 feet deep. Unlike the very old meteorite craters of the moon, however, the craters on earth are of relatively recent date. The strong forces of erosion present on earth have long since eroded and leveled all the old meteorite craters, which probably pocked the surface thickly in the earth's early period.

But if Mars formerly had a denser atmosphere and a hydrosphere, it is conceivable that at some time in its evolution it also had a biophase. Perhaps the first astronauts to land there will discover petrified organisms in the rocks now exposed on the surface of the Martian desert. Mars might today be in the stage of final desert. If that should prove to be so, we have in Mars a forecast of the earth's future.

What are the indications that the desert planet formerly had a denser atmosphere and a hydrosphere? The most important clue is the composition of the Martian atmosphere itself. Earth's atmosphere contains only about 0.03 per cent carbon dioxide; the atmosphere of Mars today is about 90 per cent carbon dioxide. But no planet could have vented so much carbon dioxide from its rocks without exuding even more water. Hydrogen, it will be recalled, is the most common element that goes into the make-up of a planet. Assuming that the genesis of all the inner planets was more or less similar, there ought to be more hydrogen combined with oxygen—that is, water—than carbon combined with oxygen—carbon dioxide—present in a planet's composition. The huge oceans of earth and the enormous amounts of water vapor in our atmosphere, as against the relatively small amounts of carbon dioxide, are clear evidence of the ratio. The gaseous products of the volcanoes on earth consist of 80 per cent water vapor, 10 per cent carbon dioxide, and 10 per cent other gases. Like the earth, Mars probably vented the greater part of its atmosphere during its planetary youth, as its mass increasingly condensed. But it is likely that some gases are still issuing from its interior. The requisite vents are present, in the form of the nineteen enormous volcanoes that can be seen in Mariner 9's spectacular photographs. The greatest of them, the Nix Olympica, constitutes the biggest volcanic mass known in our solar system. This circular volcano has a diameter of

300 miles at its base, is 75,000 feet high, and has a crater opening 35 miles across.

White banks of clouds frequently hang on the lee side of these huge volcanoes. Some scientists believe that the clouds consist of atmospheric water vapor carried aloft by updrafts along the outer slopes of the volcanoes; the vapor cools in the cold Martian atmosphere and freezes into clouds of ice particles. Similar clouds form on earth on the lee side of mountains, they point out. However, it seems more likely that the clouds consist of water vapor that is pouring directly out of the volcanoes. The fact that the interior of a crater is often filled with clouds also would appear to favor such an interpretation.

The volcanoes of Mars are probably still serving the function of gigantic chimneys, and the white clouds of steam show that the fires in the planet's interior have not yet been extinguished.

Other highly interesting details of its surface also suggest that Mars is still a highly active planet geologically. Certain diagonal structures reminiscent of tectonic tilt blocks run from southeast to northwest toward the equator, and huge faults extending through the high, dark continental mass in the southern hemisphere indicate that the crust of Mars is in constant movement to this day. From what can be seen of their tectonic structure, the southern continents of Mars seem to be drifting northward. This largely coherent land mass appears to correspond to the Gondwana protocontinent that once occupied the Southern Hemisphere of earth, until it split asunder in the Mesozoic era and its parts subsequently drifted northward.

Perhaps the most interesting geological feature of all is the enormous fault that runs like a huge gaping wound east and west across the southern continent of Mars for about 2,000 miles, and that seems to bear a genetic relationship to the diagonal structures. The circumference of the red planet is more than 13,000 miles, and the fault spans almost a sixth of it. While one end of it is in the depths of the icy night, the other end is being warmed by the sun. A temperature differential inconceivable on earth must exist between the two halves of this canyon, which is 4½ miles deep and 140 miles wide. Morning and evening, the abysses of the canyon would be whipped by sand and dust storms from one end to the other. It is possible that the wind velocity reaches 280 miles an hour. Compared to that, the worst of Saharan sandstorms would feel like a mere breeze.

There is no counterpart on earth to this vast fault line on Mars. In terms of measurements, it can be compared with the enormous fault, mentioned earlier, that extends from the Dead Sea, in Israel, down the Red Sea, through the Afar basin in Ethiopia, and across East Africa to the southern part of the continent, a distance of some 3,700 miles. Along the fault, convection currents of viscous magma rising from the interior of the earth are breaking apart the Afro-Arabian continental block and forcing the plates so far apart that a new ocean is gradually forming. In the northern part of the fault, the waters of the Indian Ocean have already flowed in to create the Red Sea. The fault is flanked by strikingly large volcanoes, especially at its southern end. The same is true of Mars. Near the western end of the great fault line are several giant volcanoes.

The fact that this 4½-mile-deep trench was not long ago filled with water is a further indication that there is no liquid water on Mars—or, rather, that there no longer is any. Geological structures comparable to the great fault can be distinguished elsewhere on the surface of Mars. The fractures that cut across the low-lying desert basins now filled with reddish dust are reminiscent of the long trenches at the bottom of the earth's oceans, such as the Middle-Atlantic trench system.

There can be little doubt that Schiaparelli, when he discovered his *canali* in 1877, was looking at these fault lines, which, as on earth, indicate active crustal movement on a planet. Schiaparelli's conjecture that the *canali* were geological in origin was proved correct, almost a century later, by scientific methods that the Italian astronomer could not have dared dream of.

We can assume that a planet like Mars, with a still active interior, has probably vented from its rocks an atmosphere corresponding in density to its size, and a hydrosphere as well. That raises the question of what has happened to important components of the atmosphere, principally its water vapor, since almost nothing but carbon dioxide is left. Perhaps the question can be answered this way: much of the atmosphere has vanished into space, and much of it has been absorbed by the rocks on the surface of the desert.

The evolution of the Martian atmosphere probably resembled that of earth. Shortly after the planet came into being, it was most likely surrounded by a protoatmosphere of reducing gases, such as methane and ammonia. Also, Mars must have exuded great quantities of water vapor and carbon dioxide. The planet had sufficient gravity to keep all the gases except the light hydrogen from escaping into space, as all the gases of our moon did. Because of the planet's distance from the sun, its atmosphere could not warm up enough to accelerate the heavy molecules of gases other than hydrogen to the requisite escape velocity. It may well be that in the early stages of its planetary evolution, perhaps for the first 2 billion or 3 billion years, when its internal activity was incomparably greater than it is at present, Mars was a planet distinctly hospitable to life. Because of the planet's greater distance from the sun, the cooling and condensation of water vapor would have taken place more rapidly than on earth. Heavy cloudbursts would have helped to form the surface of Mars and to wear away the scars of the innumerable meteorites crashing down on the planet.

For a geologically brief time span, oceans probably formed in the basins that are now filled with dust. There is no plausible reason why life could not have developed in the shelter of these primeval oceans as it did in the oceans of earth. But not until we have conducted extensive studies of the red planet's rocks will we be able to determine whether the Martian life forms succeeded in making that most important discovery for the further evolution of life—photosynthesis or a similar process that releases oxygen. However, even if we assume that some life forms on Mars did develop a process comparable to photosynthesis, the course of evolution on the two planets would have diverged at this point. No matter how much oxygen was released, a biologically determined, oxygen-rich atmosphere probably could never have formed on Mars. It is unlikely that the oxygen content of the Martian atmosphere would ever have risen to a level sufficient to protect living organisms from ultraviolet radiation, so that they could settle the dry land. Assuming that a biophase ever was initiated on Mars, for very specific reasons it would have had to be limited to the oceans.

Those reasons stem from the size of the planet. The earth had sufficient gravity to form an iron core. The gravity of Mars was not enough for that. On Mars, a larger proportion of iron in relation to the planet's mass lies closer to the surface. The presence of all that iron would have prevented the formation of a dense, oxygen-rich atmosphere favorable to life.

Water vapor molecules on Mars were split into oxygen and hydrogen by the ultraviolet portion of solar radiation. While the light hydrogen escaped into space, the highly reactive oxygen promptly combined with the iron on the surface to form iron oxides. As we have seen, those same processes probably took place during the early history of the earth. But on earth there was proportionately less iron on the surface, so that a certain amount of oxygen was left over to enrich the atmosphere. That oxygen, by blocking ultraviolet radiation, prevented the further dissociation of water vapor and of liquid water in the seas. Although much oxygen was consumed by the rocks, a sufficient level of oxygen remained to permit plants to begin producing their own organic oxygen.

That did not happen on Mars. The amount of iron on the planet's surface was so enormous that any oxygen atoms released into the atmosphere were almost immediately consumed again. There was probably not time for a self-regulating process to get started that would have shielded the water from ultraviolet radiation. As a result, all the water vapor present in the atmosphere of Mars, and gradually all the water in the seas, was dissociated. Even had there been marine organisms practicing photosynthesis and releasing oxygen, they could probably not have kept pace with the absorption of oxygen by the planet's iron surface. As enormous quantities of hydrogen escaped into space, the entire iron crust of Mars was gradually oxidized by the leftover oxygen. And that was by no means the end of the chemical process. The iron oxides themselves became great consumers of water, taking it in and becoming iron hydroxides—in other words, they rusted.

The destruction of the hydrosphere also reduced the planet's ability to store heat. Temperatures on its surface kept dropping, and the water vapor that emerged through fissures from the planet's interior probably froze before it reached the surface.

If an observer had been there to watch Mars from the very beginning, he would have seen the entire planet gradually redden from the processes of oxidation and inexorably develop into an icy, arid desert with an oxygenless atmosphere that could hold only slight amounts of water vapor. Even the hot steam that is still escaping into the red planet's atmosphere from volcanoes is immediately dissociated by ultraviolet radiation—except for what is briefly trapped in the polar regions by freezing. From data radioed to earth by Mariner 9, it has been calculated that the amount of hydrogen escaping into space every day from the atmosphere of Mars corresponds to that released by the dissociation of a million pints of water.

But what became of the hypothetical life forms of the Martian seas as those seas were slowly broken down into hydrogen and oxygen? If any sort of higher life had ever evolved on Mars, it would have had absolutely no chance of survival once the water of the sheltering seas was consumed. Only primitive life forms, mainly resistant microörganisms, could have managed to adapt to the new situation, by becoming desert dwellers rather than inhabitants of the seas. Earth provides examples of microörganisms that are not dependent on the breathing of atmospheric oxygen. It is conceivable that such life forms, staying on in the ocean basins that were gradually filling with red desert dust, could in the course of time have evolved into more complex organisms. But they could not have done without water and would have needed effective protection from ultraviolet radiation, which otherwise would have destroyed the molecular bonds within their cells.

Conditions on Mars, however, do offer a possible solution to the dilemma. Iron hydroxides—in other words, rust—absorb the blue and ultraviolet light of the sun and reflect the red portion of sunlight. (This is why we see rust as red.) Martian organisms might have found shelter from the deadly radiation under the thick layer of dust, coming to the surface whenever the dust filter is carried up into the atmosphere by those storms that sometimes last for months. Another possibility is that Martian organisms—like the many animals on earth that make shells of lime—might have developed an armor of rusty red iron hydroxide, which would furnish enough protection so that they could crawl about on the sunlit surface of the desert. The organisms would be indistinguishable in color from the surface of the desert. They might be compared to the black manganese locust, which lives in a rocky desert that is coated with manganese hydroxides.

The greatest problem for any desert organisms on Mars would be water. The Martian desert is so dry that the most arid region on earth seems wet by comparison. Could Martian organisms have developed methods for utilizing the water chemically bound in rust? But it is hard to imagine that any life forms could produce the high energy required for such decomposition. Probably they would resort to rather more terrestrial methods of obtaining water.

The scientists could hardly believe their eyes when they saw the new photographs of the Martian surface that a computer had just assembled out of the hundreds of thousands of signals radioed back to earth by Mariner 9. Yet it was undeniable: the photos distinctly showed rivers—in fact, whole river systems—that curved in broad meanders through deep clefts in the desert surface of the planet. Although no water could be distinguished in them, many indications—several tributaries leading into the main beds, for instance— suggested that these markings on the planet's surface had been caused by water erosion, not by the flow of streams of lava, say. The windings of the Martian rivers through the desert also indicated that they must be very ancient.

These dry riverbeds have one particularly interesting feature. The beds are 2½ to 3 miles wide, and their bottoms are furrowed by other, narrower channels, as though several rivers were flowing in a common bed. That is a phenomenon characteristic of desert rivers on earth, which carry varying amounts of water at highly irregular intervals. If a succession of heavy, long-lasting cloudbursts occurs in the desert, a stream of rushing water quickly fills the full width of the riverbed. If there is only a short downpour, the water flows in a small, vigorous brook that digs its own narrow channel inside the broad riverbed filled with sand and scree.

The similarity suggests that the riverbeds of Mars might owe their origin to conditions somewhat like those that have produced the desert riverbeds on earth: infrequent rainfalls varying greatly in intensity. But cloudbursts would seem to be utterly impossible under present atmospheric conditions on Mars.

At first one is tempted to assume that these must be "fossil" riverbeds left from a time when the atmosphere of Mars was much denser than it is today and when there was rain water to flow in rivers to the ancient oceans. The presence of the narrower channels inside the wide beds might perhaps be explained in terms of the planet's evolution: steadily diminishing rainfall, due to the destruction of the hydrosphere by ultraviolet light. But I cannot believe that the riverbeds date back to the planet's early period and that they have been preserved for 2 billion or 3 billion years. Though the forces of weathering are less active in a thin atmosphere like that of Mars, any such depressions would long ago have been filled with desert dust and leveled off.

As I see it, the sensational photos of Mariner 9 permit only one interpretation: at irregular intervals, and with varying intensity, devastating floods still occur on the desert planet, and quantities of water pour into the riverbeds. It could be that at times of especially violent volcanic activity large amounts of water vapor are produced, from water that failed to reach the planet's cold surface during earlier

49. Two pictures showing how the morning sun produces early-morning fog on Mars *(top, see arrows)* and, a half hour later, recondensation of water vapor in the cold air above the surface *(bottom)*

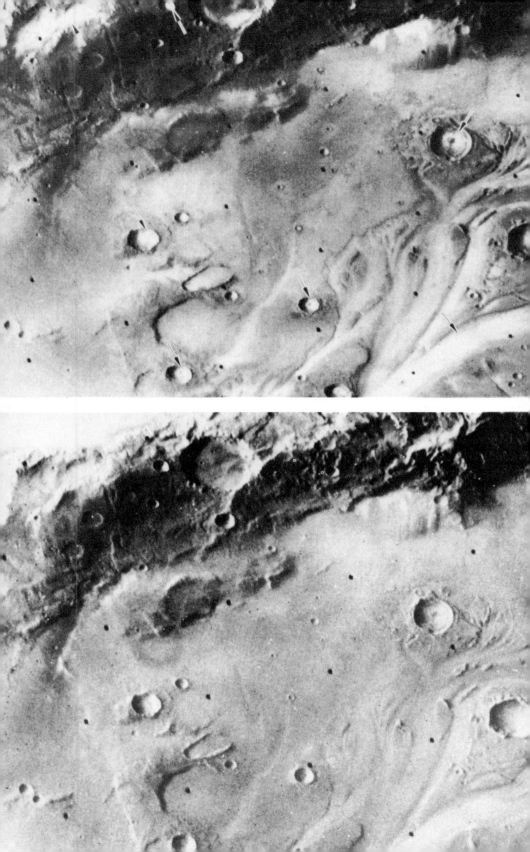

50. Crater Yuty was probably made by the collision of a meteorite with the surface of Mars. Its impact may have melted permafrost and caused the ejection of rocks.

51. Dune field on Mars with features remarkably similar to many seen in the deserts of the earth

52. Mosaic of Mars pictures of eastern part of the Chryse region show braided channels forming "tear drops"—a record of water flowing on the planet in the past.

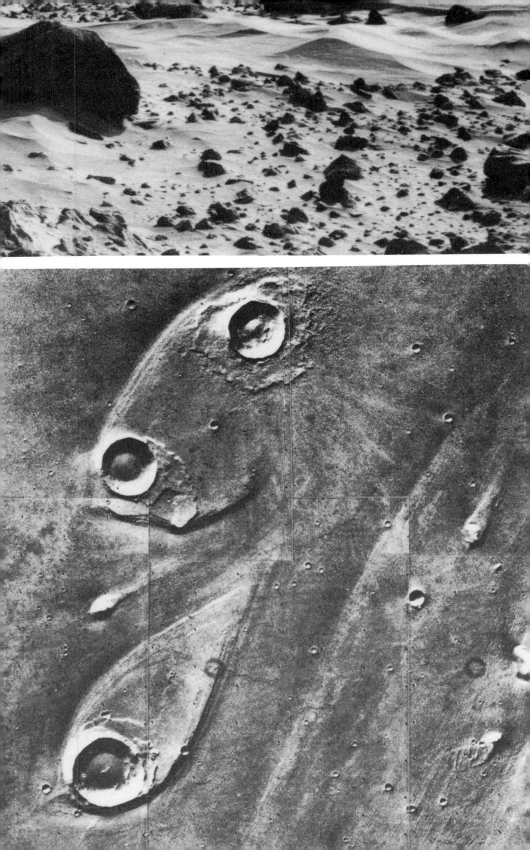

periods of volcanic eruption and froze to ice inside the volcano. Such eruptions might temporarily enrich the Martian atmosphere with so much water vapor that the dissociation of water molecules by ultraviolet light could not keep pace. The atmospheric pressure would increase, and the water vapor would cool in the cold Martian air and fall as rain on the desert, collecting in the beds of the dry rivers.

The melting of ice by volcanic activity can be observed on earth, on islands that have both glaciers and volcanoes. Iceland has several large volcanoes covered with thick masses of glacier ice, which was laid down in volcanically inactive periods. When a new volcanic eruption comes, a vast quantity of ice can melt in a remarkably short time. Riverbeds that seldom carry water, or carry little of it, become raging torrents overnight.

Other events that could produce floods of water on Mars might be changes in the planet's axis of rotation—which take place on a 50,000-year cycle—or changes in its orbit—which bring Mars closer to the sun at intervals of 95,000 to 2 million years. Either factor might at certain times lead to a warming of the planet, which would lead to the melting of water frozen on its surface and in the polar icecaps.

In any of these hypothetical situations, water would flow through the rivers and into the ocean basins now choked with dust—but not nearly enough of it to fill the ancient basins. The water would probably be soaked up by the dust.

I conjecture that any living organisms that might at one time have existed in the oceans would have withdrawn to the mouths of the Martian rivers, which presumably receive some moisture occasionally. Only there, as I see it, could they conceivably have solved the problem of obtaining water. The temporarily denser atmosphere would have provided them with additional protection from ultraviolet radiation. These last survivors of the planetary biophase on Mars would have had to span the tens of thousands, or even millions, of years between the brief periods hospitable to life by retreating into a death-like estivation. It is hard to believe that dormancy can be prolonged over such an immense span of time, but the fact is that it can. In the next chapter, we shall see that life forms on earth were capable of such a feat when their biosphere gradually turned to total desert.

On August 20, 1975, the unmanned American space probe Viking 1 started on its spectacular flight to the red planet. It was followed by Viking 2, which was launched on September 9. Almost a year later, after a flight of nearly 465 million miles, the probe swung into a Mars orbit in June 1976. The new satellite of Mars at once began taking pictures and radioing them back to earth, where they evoked both enthusiasm and alarm from the scientists and technicians at the Pasadena control center. With amazing sharpness, the pictures revealed a planet even wilder and more fissured than it had appeared in the Mariner 9 photographs. The technicians, in particular, nervously wondered whether they would be able to find a spot on that rough surface where their space laboratory could make a safe soft landing.

One overriding fact impressed the scientists: wherever they looked in the pictures of the dry Martian surface, they saw evidence that the landscape had been sculpted by the effects of running water. The water seemed to have coursed through innumerable vast river systems and to have shaped the contours of islands; but nowhere was there any sign of it in liquid form.

One of the most interesting details was revealed by photographs of the same region taken at different times of the day. As soon as the morning sun warmed the surface of the desert, water stored underground in the form of ice began to evaporate. The pictures clearly showed that the water vapor that rose by day collected toward evening, in the form of a mist of ice particles, at the lowest places on the surface, the bottoms of the numerous meteor craters (see Plate 49). Many of the craters blasted by meteorites were surrounded not by the characteristic pattern of stones and boulders hurled out by the impact, but by thick congealed slurries of mud and rock that poured out in all directions (see Plate 50). We can deduce that the meteorites falling on the planet and converting their kinetic energy into heat upon impact would have melted large quantities of ice in the subsurface. Devastating floods of water, carrying along thousands upon thousands of tons of sand and rock, must have poured out into the desert from the sites of these impacts.

One of the tasks of the Viking probes was to conduct photographic reconnaissance from orbit in order to find a suitable landing site. After studying all the photographs and measurements, the scientists decided not to let the Viking 1 lander touch down at the predetermined site, which had turned out to be too uneven. Instead, a new site was picked 560 miles away, on the western rim of the Chryse Basin.

On the morning of July 20, 1976, a radio signal separated the lander with its automatic laboratory from the orbiter. While the orbiter continued to circle the planet and survey it with its cameras and instruments, the landing vehicle dropped lower and flew over several huge volcanoes.

Throughout the lander's flight the staff at the control center sweated out a period of almost unbearable suspense. Because of the great distance, there was no way to observe the landing as it took place. "The Viking has landed just this minute, but we won't receive the signal for another nineteen minutes," a NASA spokesman pointed out. At last the tension was relieved by the signal for a perfect soft landing. There were jubilant cheers. Thirty-five minutes later, one picture strip after another began appearing on the monitors in the control center. Before the eyes of the astounded viewers pictures formed of an extraterrestrial desert, a landscape consisting of stones, boulders, and sand dunes of various sizes. No reception committee, no little green men, no cacti. The pictures might almost have been taken in the central Sahara. A camel caravan on the horizon would not have looked out of place (see Plate 51).

After some minor technical problems with the lander, which were speedily solved, on July 28 the most exciting part of the program began: the search for possible life forms in the Martian desert. The lander was equipped with a fully automatic laboratory that could carry out biochemical studies of soil samples. The experiments showed unexpectedly forceful chemical reactions, which aroused great excitement among the scientists. No inorganic materials on earth are known to produce similar reactions. But interpretation of the initial results was very difficult, and it cannot yet be stated unequivocally whether the reactions were caused by organic processes or by inorganic processes unknown on earth.

The experiments were continued, with equally inconclusive results, when Viking 2 landed safely on Mars in September 1976. But even if the probes should not detect life on the desert planet, this would be far from proof that life on Mars does not exist or no longer exists. It is equally possible to assume that the probes merely landed in the wrong places.

The emergence of life on earth, and its survival to the present day, is undoubtedly due to our planet's optimal physical characteristics: its distance from the sun and its size. On Mars, the biophase, if there ever was one, probably ended much more quickly, due to the planet's great distance from the sun and its lesser mass, with the resultant higher content of surface iron. We might say that Mars aged faster because of unfavorable characteristics. While the earth is still enjoying a biophase, though one that seems slowly to be approaching its end on the continents, Mars has already reached the phase of final desert. Can we regard Mars, then, as a model for the future of the earth?

# 8 / The End of the Earth

Throughout this book we have seen how dependent the emergence and evolution of life are on extremely complex physical conditions. Now, at the close, I shall try to answer a few questions concerning the future of our planet.

What will happen when the free atmospheric oxygen of the earth is consumed as a result of the apparently inevitable spread of deserts? Are there other physical processes, in addition to those described so far, that will make the biophase on the continents and the presence of free oxygen in the atmosphere a relatively brief episode in our planet's history, and final desert its destiny?

One of the most interesting phenomena of the earth's geology is the constancy of its temperature. From fossil plants and animals, and from certain deposits in the crust that required special temperature conditions, it has been determined that the average surface temperature on earth has not changed significantly in the past 3 billion years. For example, the average temperature in the Precambrian era, more than a billion years ago, hovered around zero on the centigrade scale—the freezing point of water—just as it does today, with deviations up or down amounting to only a few tenths of a degree.

Even such decisive paleoclimatic events as the recurrent ice ages did not produce any significant lasting changes in the average surface temperature of the earth. The ice ages were relatively brief aberrations in the long span of the earth's story.

It should be noted in passing that a long-term constant temperature by no means rules out the thesis that the deserts are spreading. There are, as I have said, both hot and cold deserts on earth, and Mars and Venus as well. It is not increasing temperature but increasing aridity that is the cause of desert formation, and the former is a consequence—on earth, at least—of the latter.

The constancy of temperature over several billion years is a remarkable physical phenomenon that has eluded explanation. The warm, life-sustaining temperatures on the surface are due largely to the earth's exogenous source of heat, the sun. But our earth, with its hot, viscous interior, also possesses an endogenous source of heat. The internal heat arises from the slow decay of the radioactive elements that have formed part of the earth from the beginning. The geophysicist Kalervo Rankama recently calculated that the earth's store of radioactive elements producing heat as they decay has been reduced by half in the course of geological history.

The earth acquired this "fuel" at the time it came into being some 5 billion years ago, and the mass of radioactive materials can only continue to diminish. Except for a few meteorites, and particles that are brought by the solar wind, no more material is being added to the earth. Just as a stove that has been filled with coal and lighted will gradually burn out and cool down, the fire in the interior of our planet will slowly go out during the next several billion years. The maintenance of a constant temperature while the endogenous production of heat has been diminishing must, therefore, be due to a compensatory increase in exogenous heat. And the only possible supplier of the requisite energy is the sun.

The earth's endogenous heat warms the crust quite evenly, except for a few volcanically active regions. The much stronger exogenous heat of solar radiation warms the planet unevenly, because of the earth's spherical shape and its rotation. The sun's rays strike only the side of the globe turned toward the sun, and they warm the region between the tropics, where they strike vertically, most of all. The radiation of heat diminishes rapidly toward the two poles. That is, the exogenous source of heat creates much sharper temperature differences, between day and night and among the different climatic zones as well.

The slow decline of our planet's interior heat must inevitably lead, over geological ages, to a diminution in continental drift, in mountain formation, and in the creation of relief deserts. A balancing of the temperature loss by an increase in solar radiation must likewise bring about significant changes in the earth's surface, and especially changes in desert formation.

But what forces exist that could influence solar radiation in such a way that it compensated for the earth's diminishing endogenous production of heat? In our search for an answer, we turn first to the moon. Is it possible that that vast extraterrestrial desert can have a palpable effect upon the temperatures of our earth?

## Tidal Friction

In 1966, the British desertologist R. F. E. Waite Peel, in a lecture on desert formation, remarked that a description of the absolute desert landscape would have to wait for the first men to set foot on the moon. Since then, men have landed on the moon and have confirmed what Peel and many other scientists had surmised—that the surface of the moon is a total desert. There could be nothing more lifeless than the surface of this satellite covered with meteorite dust and pocked with craters. The moon has no atmosphere, and neither water nor any form of life is to be found on its surface. When the moon condensed out of cosmic matter 4.5 or 5 billion years ago, its mass was too small for it to produce an atmosphere and a hydrosphere sufficient to foster life. For the same reason, its gravity was so low that it could not retain the gases vented from its rocks. From the start, the moon was condemned to remain a desert forever; it is inconceivable that at any time in its evolution it could have had even the briefest of biophases.

On the other hand, the hopeful forecast of the astrophysicist Harold Urey—"Give me a piece of the moon and I'll tell you how the solar system originated"—remains unfulfilled. In spite of enormous strides in lunar research, we cannot yet say how the earth came to acquire a moon that is unusually large in comparison with the satellites of the other planets in our solar system. The most likely hypothesis would seem to be that the earth and its satellite arose simultaneously, as a kind of double planet. If so, such a planetary marriage would of necessity have remarkable consequences for the larger of the two partners, the one inhabited by life.

For, as we shall see, the fate of the two heavenly bodies is as inseparably linked as that of a pair of Siamese twins. If the earth had no moon, or if its moon were either smaller or larger, nothing on earth would be as we know it.

But let us return to the question of whether the moon has any effect on temperatures on earth.

We know that the moon influences the tides; but only a few specialists are aware of the implications that has for our planet's past and future. I refer to the slow braking of the earth's rotation due to the gravitational force of the moon. The mechanism of this lunar brake, which at some remote date in the future could bring the rotating earth to a standstill, is called tidal friction. Just as the earth's gravitation holds the moon in its orbit, so the moon's gravitation

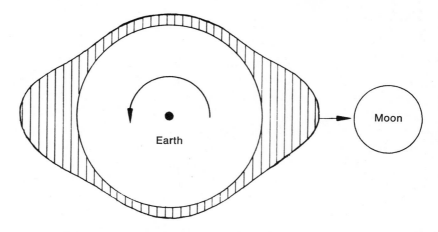

Lunar origin of the two tidal bulges

affects the earth. When the moon is over any given segment of the earth's surface, everything there is somewhat lighter than it would be if its weight were affected by terrestrial gravitation alone. The moon exerts only a small attractive force, because of its slight mass in relation to that of the earth and because of its distance from the earth, and so we do not perceptibly lose weight when the moon is overhead. Yet the attraction of the moon is strong enough to produce massive changes in the earth's surface. The most familiar and most important of those changes is the generation of two tidal bulges in the oceans on opposite sides of the earth. In addition, the moon produces tides even in the seemingly solid crust of the earth. When the moon tugs at the continents, they are raised, in a wavelike movement, as much as 6 inches.

It is not hard to understand how the tidal bulge can be created on the side of the earth facing the moon. But the existence of the second, and apparently paradoxical, tidal bulge on the half of the earth facing away from the moon is baffling, even to those who are aware of it as the cause of the six-hour alternation between low and high tide. In fact, the physical processes involved are not simple. The moon attracts not just the water on the side of the earth facing it; it attracts the entire globe, including the masses of water on the opposite side of the earth. However, the moon's gravitational attraction operates over a greater distance in reaching the other side of the earth— greater by the earth's diameter of almost 8,000 miles. Since gravitational attraction varies inversely as the *square* of the distance, the force the moon exerts on the far side of the globe is much weaker than the force it exerts on the near side or on the solid globe itself. As the globe is drawn toward the moon, the water on the side away from the moon lags behind, and that lag produces the second tidal bulge.

Since we normally regard the globe as the fixed reference point for all our activities, we are subject to many illusions about natural phenomena. We think of the sun as rising and setting, even though we know that the earth's rotation from west to east produces this event. We are similarly deceived when we stand on the ocean shore and watch the tide coming in. To us, the tidal wave appears to run around an earth that feels as if it were standing still, and we see the tide as moving toward the shore. In reality, both of the tidal bulges remain fixed in their orientation toward the moon, held there by the moon's gravitational attraction. It is the earth that turns beneath them.

That is not to say that the water in the oceans is held fast by the gravitational attraction of the moon. The same as everything else on the surface of the earth, the waters of the oceans, lakes, and rivers are subject to the earth's rotation and respond to it. Like ordinary waves stirred by the wind, those two tidal bulges fixed in their orientation by the moon consist of ever changing molecules of water. We all have at one time or another observed a piece of driftwood bobbing on the waves of a lake or ocean. If we had looked closely, we would have noted that, even in heavy seas, the object hardly moved from its position. What this tells us is that within a wave no very large mass of water is being transported from one place to another, as we tend to assume. The driftwood is merely moved up and down. The same happens to the molecules of water. Inside a wave, the molecules fall into a circular up-and-down movement. But each molecule of water transfers its kinetic energy to a neighboring molecule, initiating a kind of chain reaction of movement, which to our eyes looks like a wave moving over the surface of the water. If every wave were actually a coherent mass of water moving horizontally, no shipping would be possible on the seas.

When molecules of water cannot continue transferring their motion to other molecules, which is what happens with a wave near shore, the wave collapses; then, as anyone knows who has been roughly dashed on shore by a heavy surf, the masses of water are actually moving forward.

Although the waves stirred by the wind and the tides caused by the gravitational attraction of the moon are subject to the same physical laws of motion, the origin of their kinetic energy is decisively different. Wind-whipped waves proceed toward the coasts, and the kinetic energy they manifest as they break into surf derives from the internal, closed circulation of the earth's own energy. The two tidal bulges, on the other hand, are maintained by an extraterrestrial force, and, as I have said, they stay where they are. When the rotation of the earth brings the edge of a continent near one of the tidal bulges, its coast encounters a mound of water being held in place by the moon. The collision between a continent and a wind-driven wave merely reduces the kinetic energy of the wave; but the clashes between the continents and the fixed tidal bulges are at the expense of the earth's rotational energy. The rotation of the earth is slightly retarded by the tidal bulges. The concept of an earth turning in empty space without encountering frictional resistance is false. The earth constantly experiences friction with the tidal bulges produced by its own moon, and its angular momentum is gradually being used up over geological ages. When it came into being, the earth acquired a specific rotational energy, which accounts for its rotation around its axis. But, like a top after it has been set spinning, the earth can only lose rotational energy, because of the tidal friction; there is no way it can acquire more.

In proportion to the total mass of the earth, the water of the oceans, vast though it seems to us, covers our planet like the thinnest of films. The ratio between all the surface water and the mass of the earth is approximately 1 to 4 million. Obviously, the tidal bulges formed in this thin stratum of water can have only the slightest braking effect on the rotational energy of the massive earth. Since the beginning of our chronology nearly 2,000 years ago, tidal friction has slowed the earth's rotation by no more than a few seconds. Yet the lunar brake does make each day a tiny fraction of a second longer than the preceding day. The daily prolongations are so small that they can be calculated but never actually experienced. Yet, minuscule though they seem in terms of human history, they assume great importance in terms of geological ages. Since the speed at which the earth orbits the sun apparently does not change, the length of the year remains unchanged. But the days grow longer as the earth's rotation is braked, and this means that fewer and fewer of the longer days will fit into a single revolution of the earth around the sun. In fact, we have tangible evidence that the earth once rotated faster, the day was shorter, and there were more days in the year.

It will be recalled that in chapter 2 I described the annual rings in the Devonian corals, which show that 380 million to 400 million years ago the year had 395 days of about twenty-two hours. Around 440 million years ago, in the Silurian period, when the first plants were leaving the shelter of the oceans and advancing onto the protocontinents, the earth would have rotated around its axis in twenty-one hours and a year would have had some 410 days. Perhaps 800 million years in the future, the earth's rotation will have been braked sufficiently for a day to be some thirty hours long and the year will have correspondingly fewer days. By that distant period, the earth's endogenous production of heat will probably have declined to such a point that the movement of the crust, and the resultant cycle of mountain formation, will slowly have come to a standstill.

When that happens, the forces of erosion will gradually regain the upper hand. Although by then the mountains on the edges of the high, plateaulike, arid continents will have reached heights of 30,000 to 40,000 feet, they will be worn down again. With the geologically dead mountain barriers leveled, the rain clouds from the oceans will find it easier to reach the parched continents. That could mean the beginning of another moist phase, one far more favorable to life on land.

But, whatever the chances on earth 800 million years from now, one thing seems certain: human beings will not be around to witness them. By that period in the earth's history, our species will almost surely have been extinct for more than 700 million years. *Homo sapiens* will long since have become the victim of his own high evolutionary status and overspecialization. A general rule of evolution, applicable to all the species on earth, is that the higher they are on the evolutionary scale and the more differentiated they have become, the more sensitively they react to changes in their environment, the less they are able to adapt, and the shorter is the period of their earthly existence. There is nothing to suggest that man alone will be an exception to that rule.

Those members of our species who have not fallen victim to the clever death-dealing implements they themselves have developed, such as nuclear weapons, will be destroyed by shortages of food, the spread of deserts, and the resultant lack of oxygen in the air. In spite of the leveling of the mountain belts that keep rain clouds from the continents, the earth's future prospects are anything but alluring. The moon brake will continue to slow the earth's rotation, making the day longer and longer. Tidal friction will ultimately cancel out the possibility of a new biophase. For an increase in the length of the day will have grave consequences for climatic conditions on earth. Its sunlit half will receive more heat, but during the correspondingly longer nights the earth will have more time to radiate the heat into space. While the total solar energy received by the earth may remain the same, the temperature difference between day and night will gradually become more extreme. That, in turn, will lead to increasingly violent movements of air. The hot air masses on the sunlit half of the earth will steadily stream at high speed into the cold night side.

In the geologically remote future, sand and dust storms—hurricanes, rather—will rage constantly around the entire globe. Like the Sahara with its sandstorms and Mars with its dust storms, the earth will be subject to such vicious sandblasting that it will effectively prevent any resettlement of plants on the desert continents. Just as today the abrasive desert wind carries away old sediments, rearranges them, and uncovers the petrified remains of past life forms, in that remote future the sandblasting jets ranging around a languidly rotating earth will wear away the strata of rock in which the petrified remains of our own species, our own civilization, lie embedded. Along with innumerable petrified leg, arm, and skull bones, vast numbers of our tools will come to light—for example, segments of computers and portions of the industrial plants that today are contributing to the destruction of the earth's atmosphere. The most common fossils undoubtedly will include the remnants of our weapons.

If, in that remote future, intelligent extraterrestrial beings should land on earth to find out whether the continents of the desert planet ever went through a biophase, the multitude of fossils and traces from the Age of Man would give them a great deal to ponder. When, in rock strata made of solidified mud, they find two lines that always run parallel, they will surely be able to identify them as the tracks of automobiles, whose bodies will also be present in great numbers, if not as petrifactions, at least as imprints in the rock. No doubt there will be hot arguments among those extraterrestrial paleontologists and biologists over whether the four-wheeled remains are those of an independent life form that might once have dominated the earth. Some of the biologists may infer that the earthly life forms had succeeded in developing the highly advantageous living wheel and that the many bones of arms, legs, and skulls were the remains of organisms that had served as food for the four-wheeled monarchs of the planet.

The origin and significance of another type of imprint, found in enormous quantities in the rock strata of the future, would probably confront the extraterrestrial scientists with an even more perplexing problem. This would be a print in the shape of a ring about three-quarters of an inch in diameter. On one side of the narrow, ring-shaped imprint they would make out a small round depression from which extended a sort of stem nearly an inch long, slightly wider and neatly rounded at the end (see page 294).

In the distant future, then, the revealing forces of erosion will again turn the earth into a garbage planet.

In about 3.5 billion years, the braking effect of the tidal friction will have gone so far that the rotation of the earth around its axis will coincide with the tidal bulges maintained by the moon. Then the earth will always show the same face to the moon, and one rotation of the earth—a day—will take as long as a revolution of the moon—a month. During these extremely prolonged earth days—only twelve of them will fit into a year—the sun will shine steadily on a given portion of the earth for two weeks, and nights will likewise last two weeks. Because of the extreme heating of the day side, the corresponding cooling of the night side, and the resultant atmospheric turbulence, the climate on this doomed earth will be so forbidding that any highly developed life forms that still survive—despite the aridity and the lack of oxygen—will face insuperable problems of adjustment. Such primitive life forms as microörganisms, however, would be quite able to adjust to such a climate.

But some scientists hold that the earth will never reach that final state. They presume that before then our planet would have been destroyed by an entirely different cosmic catastrophe.


## "Some say the world will end in fire . . ."

Until recently, many astronomers assumed that the planets were expelled from a central mass, the sun, in a gaseous state. Today, astrophysicists are increasingly coming around to the view that the planets were formed from cold materials, from whirling clouds of meteorite dust and innumerable fragments of rock that were captured by the gravitation of the sun and forced into an orbit around it. All this was discussed at greater length in chapter 2.

The late German geologist Hans Georg Wunderlich* held this

---

* *Das neue Bild der Erde*, Hamburg, 1975.

Some hypothetical future fossils: tabs from beverage cans

view, and he also argued that the planets are gradually approaching closer to the sun. He contended that, in addition to the lunar brake, there is a kind of magnetic brake that has predetermined the future fate of the earth, and of the other planets as well.

In our earth's revolution around the sun, its magnetic field is constantly moving within the sun's enormous magnetic field. In Wunderlich's view, the lines of force of the two magnetospheres intersect, so that currents are continually being induced. But such induction costs energy, which is withdrawn from the earth's orbital energy. The result is that the earth's revolution around the sun is slowly being braked over vast intervals of time. The gradual consumption of the earth's orbital energy by this magnetic brake means that the centrifugal force with which our planet opposes the sun's attraction is also diminishing steadily. Ultimately, the gravitational attraction of the sun will prevail.

The effects of the lunar brake, or tidal friction, do not adequately explain the constancy of surface temperatures on earth. The slowing of rotation does not increase the total solar radiation received by the earth, but merely intensifies the temperature differential between day and night. But a closer approach to the sun resulting from the action of the magnetic brake could produce a genuine increase in temperature that would balance out the loss from the earth's diminishing endogenous production of heat.

If the Wunderlich theory is correct, the earth's orbit around the sun is not an immutable ellipse but, rather, an ever narrowing elliptic spiral terminating in the sun. Therefore, in the remote future, several billion years from now—possibly before the lunar brake could bring the earth's rotation around its axis to a standstill—the earth will plunge into the sun and burn up. Professor Wunderlich then posed the question whether that might not have happened to other planets of our solar system in the past. He suggested that several planets that might once have revolved around the sun inside the orbit of Mercury could already have plunged into our central star. Mercury is now the planet closest to the sun, and its orbit displays certain irregularities that are not adequately explained by the known laws of celestial mechanics. It would be possible for the deviations in the orbit of Mercury to have been caused by the gravitation of another planet circling between Mercury and the sun. But no such planet is present in our solar system—or, rather, no such planet is still present. Perhaps it, too, was slowed by the magnetic brake, and it plunged into the sun millions of years ago, leaving the observable effects of its gravitation on the orbit of its neighbor as testimony of its former existence.

As a planet gradually approached the sun, its atmosphere and hydrosphere would be destroyed by the increasing ultraviolet radiation, until all life on the planet was extinguished. When it finally plunged into the sun, the effect on the other planets, according to Wunderlich's theory, would be a brief—we are speaking of time in cosmic dimensions—cooling off, due to a temporary disturbance in the production of heat in the sun's interior. The other planets would experience an ice age.

But why would the sun's energy production diminish when a planet plunged into it? Some 99 per cent of the mass of our entire solar system is concentrated in the sun; the remaining 1 per cent is distributed among the nine planets, their moons, and the asteroid belt (see page 48). Given such a relationship, how could there be any effect on the energy economy in the interior of the sun from a planet's penetrating the sun's mass? If I throw a grain of sand into a stove filled with glowing coals, the stove's radiation of heat is not affected, and the example gives some sense of the mass of the sun in relation to the mass of one of the inner planets—Mercury, Venus, Earth, and Mars. But the example is irrelevant in another respect, for the sun cannot be compared to a stove filled with red-hot coals.

The sun is an enormous fusion reactor floating freely in space and held together by gravitational forces. In its interior, under a pressure of 200 billion tons and a temperature of 15 million degrees centigrade, energy is produced by the fusion of hydrogen nuclei into helium. Every second, 657 million tons of hydrogen in the center of the sun are fused into helium, and that process has been going on for 5 billion years. That gives some conception of the enormous size of our sun—which is a dwarf by comparison with some of the giant stars in our galaxy.

From experimental attempts to imitate the solar power plant—not only in the form of the hydrogen bomb, but also for the purpose of obtaining controlled fusion energy—we know that the fusion reaction is extremely sensitive to contamination, especially by matter whose nuclei have the atomic weight of iron or the elements close to it in the periodic table. The sun is made up of hydrogen and helium; the inner planets consist of elements with comparatively heavy nuclei. If an inner planet were to plunge into the sun, its high density would cause it to fall all the way into the center. Thus—still following Professor Wunderlich's reasoning—it would penetrate into the sun's fusion reactor and, by contaminating the "burning" hydrogen, would seriously disturb the nuclear fusion process and bring about a reduction in the production of energy.

Just as heat from the interior of the earth is conducted to the crust and in part to the surface by rising convection currents, so is the fusion energy produced in the sun's center carried to the circumference, from where it is radiated into space. The planetary contamination would probably also be moved around by the convection currents, so that it would disturb the fusion reaction for only a relatively brief period at any one time. But the disturbances would recur. As a planet that had plunged into the sun slowly burned up over the course of several million years, its shrinking mass would be carried by the convection currents in and out of the fusion reactor in the center of the sun. The contamination would produce intermittent disturbances of the internal fusion reactor, although over the long run the sun would continue to produce energy and radiate heat at a relatively even rate.

Professor Wunderlich has seen in the climatic history of the earth clear and impressive evidence that this has already happened. As I have said, at intervals of about 300 million years there have been "brief" interruptions, lasting several million years, during which the earth's generally warm climate has perceptibly cooled. Within these cool periods fluctuations have occurred between distinct ice ages and warmer interglacial periods. Each time a period of ice ages has come to an end, the climate has remained prevailingly warm for the next 300 million years.

Many explanations have been offered for the ice ages and the climatic fluctuations within them. In the past century, as scientists have gathered more and more information about the last great glaciation of the Northern Hemisphere, at least several dozen different hypotheses have been proposed. While each offers plausible explanations for certain phenomena of the ice ages or for the onset of glaciation in general, none has adequately accounted for the long-term evolution of our climate over the whole course of the earth's history. In particular, no explanation has been suggested for the exceedingly important phenomenon of our planet's constant surface temperature, in spite of ice ages and the diminishing endogenous production of heat. Professor Wunderlich's new hypothesis of a magnetic brake operating between the planets and the sun does fill that gap. His theory of the earth's gradual approach to the sun serves to explain how the loss of the earth's internal heat is balanced by increased solar radiation.

Let us for a moment conjecture, with Wunderlich, that a planet has plunged into the sun at intervals of approximately 300 million years and that the last such event took place about 2.5 million years ago. That would correspond with the beginning of the latest period of ice ages. As the contamination from the hypothetical planet was repeatedly carried by convection currents into the sun's fusion reactor, reducing its production of energy, the earth would have experienced glacial periods and interglacial warm periods, the latter probably occurring when the planetary contamination was carried back out of the central zone of the sun. The origin of man dates from this latest glacial period, with its considerable climatic fluctuations. The question is whether we are living today in one of the warm interglacial times, which would mean that the glacial period has not yet ended and that mankind perhaps faces an imminent ice age. Or has the last of the planetary contamination inside the sun been so far reduced that the era of glaciation is largely over? If the latter is the case, we can assume that the earth is already at the beginning of the next warm age, due to last some 300 million years, during which exogenous solar radiation will more and more balance out the loss of the

planet's internal heat, partly because the sun's fusion reactor is working full blast again and partly because the earth is slowly but inexorably approaching closer to the sun. In that event, the increasing
heat, along with the processes of oxygen consumption and lunar braking, will gradually transform the surface of the earth into a final
desert.

## ". . . some say in ice"

Even if Professor Wunderlich's hypothesis of the magnetic brake
operating between the earth and the sun proves false, the earth's
prospects for the future are not rosy. For there would still be the
lunar braking by tidal friction, slowing the earth's rotation until a day
was as long as a month, with climatic consequences that would devas-
tate our planet. From then on, the further thermal fate of the earth
would depend in another way on the fate of the sun. Astrophysicists
have likewise formulated some theories about that. They know how
the sun will evolve because they have examined stars in the universe
that are older than our sun.

We have already discussed the evolution of our sun and other stars
out of a gigantic cloud of cosmic dust and gas. We know that after a
number of phases of instability the contracting forces of gravity and
the expanding forces of radiation achieved an equilibrium that has
already lasted some 4.5 billion years and will probably go on for an
equal length of time.

If our sun were to continue burning "only" 657 million tons of
hydrogen every second, its supply of fuel would last another 50 billion years. But long before then, probably in another 4.5 to 5 billion
years, the sun will have converted about 15 per cent of its hydrogen
into helium by nuclear fusion; at that point it is likely to enter an
unstable phase in its evolution. As soon as the sun has consumed a
significant part of the hydrogen in its center, the equilibrium between
radiation pressure and gravitational forces will break down. A diminishing supply of fuel will lead inevitably to diminishing energy production and to an overbalance in the gravitational forces that seek to
contract the mass of the sun.

But the nuclear processes that take place in the sun's interior are so
complex that at first the effect will be just the opposite. Let me try to
simplify the exceedingly complicated interrelationships.

With the consumption of more of the hydrogen in the sun's interior, the zone of fusion first shifts outward toward the solar rim. The heavier helium ash that results from the fusion process accumulates more and more rapidly in the center of the sun, increasing the gravitational pressure toward the center. That in turn raises the temperature in the interior. In its turn, the intense heat radiating outward accelerates the burning of the remaining hydrogen. At this stage, the sun is fusing its hydrogen into heavier helium at a breath-taking pace, always increasing the quantity of the helium ash and so enormously accelerating the cycle.

As the nuclear process moves slowly toward the surface, steadily accelerating because of the constantly rising temperature, the stream of energy pouring outward increases correspondingly. As a result, the sun rapidly swells to an enormous size. A point is reached at which the zone of hydrogen fusion is so close to the surface that the pressure no longer suffices to keep the solar reactor going. The nuclear fires in the strata close to the sun's surface go out, and the outward pressure of radiation drops rapidly. Gravitational forces once more take over, and the enormously swollen sphere of the sun collapses.

By this remote time in the future, the transformation of hydrogen into helium that has been taking place inside the sun for some 10 billion years has accumulated vast quantities of helium ash. The ash, under its own weight and under the increased gravitational force resulting from the burning out of the hydrogen, contracts to such a point that the pressure in the center of the sun rises far beyond 200 billion tons and the temperature goes from an inconceivable 15 million degrees centigrade to a likewise inconceivable 110 million degrees. At that temperature, the helium itself begins to fuse and is transformed into a still heavier element, carbon. The resumption of nuclear reactions releases energy to oppose the gravitational forces and prevent the further contraction of the solar mass. The sun, transformed from a hydrogen sun into a helium sun, again enters a stable stage, although one that is destined to be brief by the dimensions of cosmic time.

Viewed from the earth, the diameter of the sun would appear about twenty times as large as it does today, and the gigantic red star would radiate so much energy that surface temperatures on earth would rise to over 500° C.—enough to bring the oceans to a boil, evaporate their water, and kill any higher life forms that might have survived to this remote era. The enormous amount of ultraviolet light emitted by the giant sun would dissociate the hot steam cloaking the globe into its component elements of hydrogen and oxygen. As all the hydrogen escaped into space, because of its high temperature, and as the last of the released oxygen was absorbed by the earth's crust, the surface of the earth would dry out completely. Although the ocean basins would fill with red desert dust, as on Mars, there would still be ponds, lakes, and small watercourses here and there on earth. However, they would be filled not with water but with molten lead, and on the hot surface sulfur would boil, as predicted in the Bible.

Yet, in spite of these apocalyptic conditions, life on earth might still have a last chance of survival. The nature of the final existing micro-örganisms on earth some 5 billion years off can, in a sense, already be observed in the deserts today.

The Great Salt Lake of Utah is the last remnant of a vast inland sea that once covered much of North America west of the Rocky Mountains. It slowly dried out as the climate of North America became more arid. The evaporated inland sea left behind 4 billion tons of minerals, chiefly salt, that had been dissolved in the sea water. These deposits on the former sea bottom have created a white, perfectly flat desert of salt that stretches from the shore of the remaining lake to the horizon.

The ancient sea bottom conveys some impression of what will happen in future geological ages when the oceans dry out. The sparse precipitation that continues to bring water to the Great Salt Lake cannot keep pace with the enormously high rate of evaporation. Ancient shorelines march in terracelike stages up the flanks of the mountains that tower above the sterile salt desert—their peaks were once islands rising out of the waters of the inland sea. Those shorelines show that the water level has already dropped more than 250 feet.

The evaporation of so much water has enormously increased the salt content of the remaining water in the Great Salt Lake. Depending on the water level—which varies from year to year with the precipitation received by the mountains bordering the lake on the east—the salt content ranges between 18 and 24 per cent. In parts of the lake that are largely cut off from inflows of fresh water, the water has a salt content of 28 per cent—virtually the saturation point. It is impossible to drown in this lake that has a salt level six to nine times higher than that of the oceans.

Because of the salt content, the Great Salt Lake is extremely hostile to most forms of life, and only a few primitive organisms inhabit its water—although these are present in astronomical numbers. Fish have long since vanished. In addition to some species of bacteria, a number of salt-resistant algae inhabit the lake, manufacturing food by photosynthesis and releasing oxygen. At certain seasons, the micro-örganisms multiply so incredibly that they color the water of the Great Salt Lake orange, green, or red, depending on the species involved.

The algae provide food for the only species of animal that has survived in the salty water. This is the tiny, primitive brine shrimp, about 10 millimeters long, which inhabits the lake by the billions.

On windless days in certain seasons of the year, it is possible to observe curiously thick, oblong clouds above the lake, particularly where the water meets the shoreline of the white, perfectly flat salt desert. The clouds move very rapidly, constantly changing color; at one moment they look black, at another silvery. The clouds consist of innumerable little flies that also form part of the biologic facies of the lake. These flies, which belong to the genus *Ephydra*, have adapted to the hot, salty environment; their distribution is limited to the vicinity of the salt lake. They feed on algae and are protected from the glaring sunlight by the silver coloring of the upper part of their bodies, which reflects a large percentage of the solar radiation. A similar color adaptation is found in another silvery insect, called the aluminum ant, which inhabits the Sahara.

The drying out of the remainder of the inland sea is proceeding slowly but inexorably, and someday the area where a blue lake extended from horizon to horizon will be only another stretch of white, completely lifeless salt desert. What will happen then to the last life forms can be observed in the parts of the lake in which the salt content of the water has reached 28 per cent. In that saline solution, the brine shrimps die first, and then the algae. When the salt begins crystallizing out of the water and the area becomes a salt desert, even the bacteria are condemned to death.

But before dying the last shrimp generation will have laid billions of highly resistant eggs. The eggs, embedded in the salt, will remain viable, giving the species a further chance at survival if unusually heavy rains should cause a temporary rise in the water level of the lake. In the long run, however, the lake itself is condemned to death, and along with it all the life forms that still exist in its waters. If you pick up a lump of salt from the surface of this desert, you can see with the naked eye the desiccated remains of shrimps in it. If you place the lump under a microscope, you can also make out the dead algae and bacteria. Someday the wind may deposit eroded sand and dust over the dry sea basin, covering the strata of salt, which will then become one more of the many stratified deposits that make up the outer portion of the earth's crust. At that point, the story of life in the salt desert would seem to have come to an end. In fact, the story has a fascinating continuation.

About fifteen years ago, the sanatorium directors at Bad Nauheim, a German spa noted for its salt baths, asked a local scientific institute to investigate one of its warm saline springs. For some time, the spring had been subject to repeated bacterial contamination, the source of which could not be determined. The Nauheim microbiologist Heinz J. Dombrowski set to work, but he soon discovered that the contamination was caused by a living bacterium of a species completely unknown to him, although he was a noted specialist in the field. Accordingly, contamination originating outside the spring seemed unlikely. The bacteria appeared to come from the spring itself or from its underground aquifer. The geological structure of the area was well known; the salt in the water of this particular Nauheim spring came from a tremendous salt deposit almost 2,000 feet deep. The salt had been laid down in the Permian period, some 250 million years ago, when the Zechstein Sea, which then covered large parts of central Europe, was repeatedly cut off from the rest of the world's oceans and its water evaporated during several cycles of temporarily arid climate. The process can be compared to the evaporation of the great inland sea that left behind the salt desert of Utah.

In their further researches, the Nauheim scientists drilled down to the Zechstein salt through which the spring water had passed and brought samples to the surface. Enclosed in the crystals of salt were dead bacteria. With no small degree of suspense, the scientists placed the crystals holding the fossil bacteria in a watery nutritive solution. They could hardly believe their eyes. The bacteria appeared to thaw out and awaken to new life—after 250 million years. They displayed an amazing vitality and ability to reproduce, and within a short time they had formed large and vigorous cultures.

To make certain that the bacterial cultures were truly resurrected from bacteria that had been enclosed in the salt 250 million years ago, and were not the result of accidental contamination, the scientists carefully sterilized each new piece of salt before taking a sample of it. The positive results were repeated. Further evidence that 250-million-year-old life forms had actually been awakened from dormancy was the fact that, out of a total of forty resurrected species of bacteria, only a single species bore any resemblance to one of the innumerable species existing today—and that resemblance was to the present-day agent of typhus. For this reason, the experiments were conducted with maximum precautions. One of the resurrected species was finally identified as the agent responsible for the contamination of the spring. After an inconceivably long period of time, the conditions that had allowed the species to persist in the drying Zechstein Sea of the Permian period were by chance restored in the saline water of the spring that brought the dormant bacteria to the surface.

But how can microörganisms survive in suspended animation for such incredible ages? Some years before Dombrowski's sensational discovery, American scientists had found one answer to the question, although they were working in an entirely different context. They were studying methods of preserving bacteria. The most reliable method, they found, was to grow the bacteria in a nutrient solution, and then slowly enrich the solution with salt, until the saturation point was reached and the salt crystallized out. The crystallization completely dehydrated the microörganisms. In that dried state, their metabolism ceased altogether and they remained viable almost indefinitely.

During such suspended animation there are none of the changes that presumably cause aging and death. The time factor vanishes from life. But—it must be emphasized—these processes are conceivable only in such primitive life forms as monocellular bacteria.

Without knowing it, the American scientists had exactly imitated the experiment conducted by nature 250 million years ago when the ancient Zechstein Sea gradually dried up, leaving behind only salt.

When the Nauheim biologists applied their American colleagues' method of preservation to the resurrected bacteria, they actually succeeded in returning the bacteria, or their progeny, to the state of suspended animation.

Similar life forms have been found in a still older salt deposit in Canada, dating from the Devonian period and estimated to be 320 million years old. These, too, were resurrected. The length of time the microörganisms have survived can perhaps be grasped if we think of them as belonging to the period when the first plants left the shelter of the oceans to settle on the protocontinents. Theirs was an epoch in which the domination of the primal desert, which had lasted several billion years on the continents, was coming to an end. The continental biophase, which now seems to be approaching its end, was just beginning.

The knowledge that certain primitive microörganisms are capable of remaining viable in a state of suspended animation for a practically unlimited time—as if they were the very idea of Life itself—has far-reaching consequences (in the most literal sense of the word far-reaching) for the future of this earth.

If, in remote times to come, our sun showers the planets with so much energy that the earth's crust becomes parched, the ocean basins will be converted into salt deserts. The salt will cover the bottom of the former seas in a stratum several hundred feet thick. By then, all the higher life forms will have become extinct, and their remains will have fallen to dust. But microörganisms that have already survived one apocalypse will be locked by the myriads in crystals of salt.

The death agony of the sun will take long ages. It will not be able to maintain its lavish emission of energy indefinitely, and after about 2 billion more years it will have transformed the greater part of its new fuel, helium, into the heavier element carbon. Once again, the outpouring of energy from the interior of the sun will diminish, contraction will set in, and this time temperatures far above 110 million degrees centigrade will be engendered. In fact, they will be high enough to consume the carbon as a nuclear fuel and transform its atoms into the still heavier element neon. Each new element that serves as nuclear fuel will provide the ashes for the next nuclear reaction; each presumably will follow the last at a progressively shorter interval. During this period of stellar evolution our sun will pass through several unstable phases. Annihilating quantities of deadly gamma rays and x-rays will pour from its interior to the surface and be radiated out into interplanetary space. Those rays will literally sterilize the surface of the earth.

At present, the greater part of the deadly radiation produced by the hydrogen nuclear reactor in the sun's interior is transformed into harmless long-wave radiation during its long journey to the solar surface. That is the chief reason the earth is not bombarded by radiation too deadly for its atmosphere to fend off, but receives, instead, light and heat.

The production of heavier and heavier elements, along with repeated phases of contraction, will cause our sun to shrink more and more. The sun's entire mass will condense to an incredible density, so that a cubic centimeter of matter from its core will weigh many tons. Small in size, radiating an intense white light, it will have become one of those stars to which astronomers have given the name white dwarfs.

The nuclear production of increasingly heavy elements will make nuclear reactions steadily harder to sustain. Astrophysicists have calculated that, for a star the mass of our sun, a point is rapidly reached at which gravitation no longer suffices to keep the reactions going in the newly created heavy elements. The heavy atomic nuclei resist further concentration, and no new nuclear fires are generated. Slowly, over billions of years, the nuclear reactions in the interior of our sun will inevitably come to a stop. By that time, the sun will have transformed only about one-thousandth of its enormous mass into energy. That is much too slight a loss of mass to break the bonds of gravitation between the central star and its planets. Shrunk to an inconceivable density, the mighty mass will continue to hold the earth and the other planets in their orbits.

In that remotest of futures, which we can no longer even begin to imagine, life on earth will have another chance, its last. On a dead and slowly cooling earth, some last remnants of water vapor will probably rise from the interior. The water vapor will again, as happened when the earth first formed, condense and collect in the dried-out basins of the former oceans, though it will not be nearly enough to refill them. The water of these newly resurrected little seas will dissolve the layers of salt and awaken the microörganisms locked within the salt crystals of the deeper strata, where they have been protected from the deadly radiation; after perhaps several billion years in suspended animation, those bacteria will come to life again.

The sun will continue to cool until the temperature on its surface reaches the temperature of space. Long before then, earth will have suffered its ultimate death, the cold death that the Swedish cosmologist Svante Arrhenius impressively described in 1926:

"If we wish to form a picture of the ultimate fate of our earth as someday it sinks into darkness and cold under a cooled and weakened sun, we should look for a model to Mars, not to our moon. Slowly the time will come when the ocean begins to freeze over, until at last it freezes to the very bottom. The rain will fall more infrequently; only light snowfalls will introduce some change into the appearance of the earth's surface. As far as the dry land extends, it will turn more and more into a sandy desert. From cracks in the ground the gases will pour out of the earth's interior and will appear as dark streaks. When the temperature at the Equator drops below the freezing point, the polar regions will be the only ones in which a light coating of frost will thaw in midsummer and where the last feeble organisms, after a long winter sleep in seed or spore, will drag

Comparative magnitudes
of various suns

1. Sun
2. Procyon
3. Altair
4. Sirius
5. Algol
6. Aldebaran
7. Pole Star
8. Rigel

out their hard life.* Ultimately even the last remnant of life will vanish, and only the sandstorms will vary the dreary monotony, and the vapors that will continue to rise out of the earth's interior from mountain crevices. Meteoric dust falling to earth, which now lies undisturbed only at the bottoms of the seas, will cover the entire surface with a mantle that oxidation will color brick-red. But once the oxygen is used up, the meteoric dust will retain its green color, furnishing earth with a shroud."

But such a cold end would not be normal for a star in the universe. In fact, the cosmos offers us another and more consoling alternative.

In many suns that possess much greater masses than our relatively small one, gravitational contraction continues steadily, far beyond the stage of a white dwarf. Under the ever higher pressures and temperatures in their interiors, more and more heavy elements are produced, each providing the fuel for the next stage of nuclear fusion. Neon ashes kindle at a temperature of 800 million degrees, leaving behind the much heavier ash of magnesium. At 1.5 billion degrees, silicon, sulfur, and phosphorus are formed. As ever heavier elements are created in the interior of the star, the nuclear fusion zones of the lighter elements move steadily outward in a succession of shells. While light helium is burning in an exterior layer, a heavy metal will be burning farther in toward the core. At this stage, the star in effect constitutes a gigantic nuclear bomb, of cosmic dimensions, and equipped with a time fuse. Ultimately, a point is reached in the interior of the star at which gravitation destroys the atomic structure of the star's constituent matter. Under pressures and temperatures that are altogether beyond human imagination, the structures of the elementary particles that make up the atomic nuclei break apart, and the star collapses upon itself. At the resultant temperature of up to 6 billion degrees—which, among other things, creates the heavy elements of titanium, manganese, iron, nickel, and copper—a titanic nuclear explosion destroys a large part of the stellar mass and hurls it outward into space. That is what happens when a supernova burns with the brightness of 300 million suns. During this process, the exploding sun, which has built up so many heavy elements from the primal matter of the cosmos, hydrogen, scatters those heavy elements out into the universe—where, at some place and some time in the infinity of the cosmos, new planets and new life can arise from them.

---

* Because of the inclination of the earth's axis, the polar regions—the Arctic in summer, the Antarctic in winter—receive continuous sunlight twenty-four hours a day.—AUTHOR'S NOTE

# Acknowledgments

This book, and the personal researches described in it, would not have been possible without my wife's active participation, even in explorations that often involved great physical strain and hardship.

I wish to thank two North African friends, Monsieur Benamar and Monsieur Mohammed Seddiqui, for their hospitality and assistance.

I am grateful to the following persons for checking certain portions of the manuscript: Dr. Gerhard Alberti, Dr. Claudia Nebelsiek, Dr. Ulrich Nebelsiek, and Dr. Peter von der Osten Sacken.

U.G.